# The Spatial
# Foundations
# of Language
# and Cognition

Edited by
KELLY S. MIX, LINDA B. SMITH,
AND MICHAEL GASSER

OXFORD
UNIVERSITY PRESS

# OXFORD

UNIVERSITY PRESS

Great Clarendon Street, Oxford OX2 6DP

Oxford University Press is a department of the University of Oxford.
It furthers the University's objective of excellence in research, scholarship,
and education by publishing worldwide in

Oxford New York

Auckland Cape Town Dar es Salaam Hong Kong Karachi
Kuala Lumpur Madrid Melbourne Mexico City Nairobi
New Delhi Shanghai Taipei Toronto

With offices in

Argentina Austria Brazil Chile Czech Republic France Greece
Guatemala Hungary Italy Japan Poland Portugal Singapore
South Korea Switzerland Thailand Turkey Ukraine Vietnam

Oxford is a registered trade mark of Oxford University Press
in the UK and in certain other countries

Published in the United States
by Oxford University Press Inc., New York

British Library Cataloguing in Publication Data

Data available

Library of Congress Cataloging in Publication Data

Data available

Typeset by SPI Publisher Services, Pondicherry, India
Printed in Great Britain
on acid-free paper by the
MPG Books Group, Bodmin and King's Lynn

ISBN 978–0–19–955324–2

3 5 7 9 10 8 6 4 2

# Contents

Section III.  Using Space to Ground Language

# Foreword: Space as Mechanism

Spatial cognition has long been a central topic of study in cognitive science. Researchers have asked how space is perceived, represented, processed, and talked about, all in an effort to understand how spatial cognition itself works. But there is another reason to ask about the relations among space, cognition, and language. There is mounting evidence that cognition is deeply embodied, built in a physical world and retaining the signature of that physical world in many fundamental processes. The physical world is a spatial world. Thus, there is not only thinking *about* space, but also thinking *through* space—using space to index memories, selectively attend to, and ground word meanings that are not explicitly about space. These two aspects of space—as content and as medium—have emerged as separate areas of research and discourse. However, there is much to be gained by considering the interplay between them, particularly how the state of the art in each literature impacts the other.

Toward that end, we have assembled chapters from a diverse group of scientists and scholars who represent a range of perspectives on space and language. They include experimental psychologists, computer scientists, roboticists, linguists, and philosophers. The book is divided into three sections. In the first, we address the notion of space as the grounding for abstract thought. This idea solves a number of problems. It explains how complex concepts without clear physical referents can be understood. It specifies how 'here-and-now' perception can interact with cognition to produce better problem solving or language comprehension. For example, Clark provides many excellent examples of ways that people co-opt both language and space to scaffold complex behavior. Due to this similarity in function, he contends, language and space are naturally coupled in human cognition. Ramscar, Matlock, and Boroditsky summarize a series of elegant experiments demonstrating that people ground their concepts of time in their own bodily movements. Likewise, Spivey, Richardson, and Zednik present research showing how people scan space as a way to improve recall. Together, these two chapters provide strong support for the basic idea of embodiment in cognition and, more specifically, the way movement through space is recruited by seemingly abstract cognitive processes. Mix's chapter looks forward—asking whether, if these ideas about human cognition are correct, they can be used to improve instruction in mathematics. She focuses on the role of concrete models, in particular, and asks whether they might engage a natural predisposition to ground abstract concepts in space and action.

Although spatial grounding provides a plausible explanation for higher-level processing, where does this conceptualization of cognition leave us with respect to spatial cognition in particular? As for many areas within cognitive psychology, spatial cognition was traditionally characterized in terms of logical abstractions. Research with adults has emphasized the use of propositions and linguistic frames for representing space. Developmental research has focused on how children move from concrete, egocentric views of space toward the abstract mental maps supposedly used by adults. In light of this, the claim that abstract cognition is anchored by space has a certain irony to it. Still, the same movement that questioned the grounding of other thought processes has led experts on spatial cognition to consider the role of embodiment there, too. The chapters in Section II address this issue head-on. Each grapples with the tension between established frameworks for spatial thought and mounting evidence for embodiment. Although all the authors admit a role for bodily experience, they differ in the extent to which they are willing to jettison, or even modify, traditional descriptions. But the debate itself raises critical questions about what representations are, what constitutes embodiment, and whether we need both to explain human behavior.

For example, Carlson focuses on the acquisition of spatial terminology, arguing that distance comes along for the ride as children learn a variety of spatial words—even those that are not about distance (e.g. 'above'). Distance, she posits, is part of the reference frame used for all spatial terms, and thus becomes incorporated incidentally. Similarly, Huttenlocher, Lourenco, and Vasilyeva argue that the way children encode spatial information varies depending on whether they are moving through space as they track a target. Thus, both accounts identify a role for movement in spatial cognition, but also contend that it contributes to some form of mental representation. Landau, O'Hearn, and Hoffman make an even stronger, explicit case that abstract representations are needed to complete spatial tasks, such as block design, based on their study of spatial deficits in children with Williams syndrome. In contrast, Lipinski, Spencer, and Samuelson question the need for such mental structures. They present a dynamic field model that shows how spatial language and memory for location could be connected without an intervening representation.

In Section III, we consider space as a mechanism for language acquisition—as the medium through which many words are learned, not just terms for space. Smith and Samuelson's chapter points out that spatial contiguity between word and sensory experience is likely just as powerful as temporal contiguity in promoting word learning, perhaps even more so because spatial contiguity can persist through time. However, for this mechanism to work, children would have to notice and encode spatial location along with other sensory information, like the sounds of a spoken word. Smith and Samuelson argue that research on the A-not-B error demonstrates that children do connect space and action, and this same process could become activated in word learning. Similarly, Yu and Ballard consider the way space unites word and referent, but instead of short-term memory, they focus on the role of attention. They present a series of experiments in which a robot is taught the names of objects in a picture book. This appears to hinge on joint attention between the robot and its teacher, such that spoken words co-occur with visual perception of their referents (i.e., the appropriate book illustrations), more frequently than not. Cannon and Cohen also consider the role of space in word learning, but focus on the extent to which bodily experiences (i.e., movements through space) support the acquisition of verb meanings. They make the critical point that language is grounded in space, even when the particular words are not about space.

# List of Plates

# List of Figures

# Notes on Contributors

DANA BALLARD is a professor of Computer Science at the University of Texas–Austin. His main research interest is in computational theories of the brain, with emphasis on human vision. With Chris Brown, he led a team that designed and built a high-speed binocular camera control system capable of simulating human eye movements. The theoretical aspects of that system were summarized in a paper, 'Animate vision', which received the Best Paper Award at the 1989 International Joint Conference on Artificial Intelligence. Currently, he is interested in pursuing this research by using model humans in virtual reality environments. In addition he is also interested in models of the brain that relate to detailed neural codes.

LERA BORODITSKY is an assistant professor of psychology at Stanford University. Her research centers on the nature of mental representation and how knowledge emerges out of the interactions of mind, world, and language. One focus has been to investigate how the languages we speak shape the ways we think.

ERIN CANNON received a BA in Psychology from University of California, Irvine, in 1998 and a Ph.D in Developmental Psychology from the University of Massachusetts, Amherst in 2007. Her research spans from infancy to the preschool ages, and focuses on the development of action and intention understanding, action prediction, and verb learning. She is currently a postdoctoral research associate at the University of Maryland.

LAURA CARLSON is currently Professor of Psychology at the University of Notre Dame. She earned her Ph.D from the University of Illinois, Urbana-Champaign in 1994, and has been at Notre Dame ever since. Her primary research interest is in spatial language and spatial cognition. She employs empirical, computational, and psychophysiological measures to investigate the way in which the objects and their spatial relations are encoded, represented, and described. She co-edited (with Emile van der Zee) the volume *Functional Features in Language and Space: Insights from Perception, Categorization, and Development*, published by Oxford University Press. She currently serves as Associate Editor for *Memory and Cognition*, and Associate Editor for *Journal of Experimental Psychology: Learning, Memory and Cognition*, and is on the editorial boards of *Perception and Psychophysics* and *Visual Cognition*.

ANDY CLARK is Professor of Logic and Metaphysics in the School of Philosophy, Psychology and Language Sciences at Edinburgh University. Previously, he was Professor of Philosophy and Cognitive Science at the University of Sussex, Professor of Philosophy and Director of the Philosophy/Neuroscience/Psychology Program at Washington University in St Louis, Missouri, and Professor of Philosophy and

Director of the Cognitive Science Program at Indiana University, Bloomington. He is the author of six books including *Being There: Putting Brain, Body And World Together Again* (MIT Press, 1997), *Natural-Born Cyborgs: Minds, Technologies And The Future Of Human Intelligence* (Oxford University Press, 2003), and *Supersizing the Mind: Embodiment, Action, and Cognitive Extension* (Oxford University Press, 2008). Current research interests include robotics and artificial life, the cognitive role of human-built structures, specialization and interactive dynamics in neural systems, and the interplay between language, thought, and action.

PAUL COHEN attended UCSD as an undergraduate, UCLA for an MA in Psychology, and Stanford University for a Ph.D in Computer Science and Psychology. He graduated from Stanford in 1983 and became an assistant professor in Computer Science at the University of Massachusetts. In 2003 he moved to USC's Information Sciences Institute, where he served as Deputy Director of the Intelligent Systems Division and Director of the Center for Research on Unexpected Events. In 2008 he joined the University of Arizona. His research is in artificial intelligence, with a specific focus on the sensorimotor foundations of human language.

MICHAEL GASSER is an associate professor of Computer Science and Cognitive Science at Indiana University. He earned a Ph.D in Applied Linguistics from the University of California, Los Angeles in 1988. His research focuses on connectionist models of language learning and the linguistic/digital divide.

JAMES E. HOFFMAN received his BA and Ph.D degrees in Psychology from the University of Illinois, Urbana/Champaign in 1970 and 1974, respectively. His research interests include visual attention, eye movements, and event-related brain potentials, as well as spatial cognition in people with Williams syndrome. He is currently a professor in the Psychology Department at the University of Delaware.

JANELLEN HUTTENLOCHER received her Ph.D from Radcliffe (now Harvard) in 1960. She has been on the faculty of the University of Chicago since 1974. Her longstanding research interests have focused on children's spatial development and on language acquisition, both syntactic and lexical development.

BARBARA LANDAU received her Ph.D degree in Psychology from the University of Pennsylvania in 1982. Her research interests include spatial representation, language learning, and the relationship between the two. She is currently the Dick and Lydia Todd Professor of Cognitive Science and Department Chair at the Johns Hopkins University in Baltimore, MD.

JOHN LIPINSKI received his BA in Psychology and English from the University of Notre Dame in 1995 and his Ph.D in Cognitive Psychology from the University of Iowa in 2006. His research focuses on linguistic and non-linguistic spatial cognition, with a special emphasis on the integration of these behaviors through dynamical systems and neural network models. He is currently a post-doctoral researcher at the Institut für Neuroinformatik at the Ruhr-Universität in Bochum, Germany.

STELLA F. LOURENCO received her Ph.D in Psychology from the University of Chicago in 2006. She is currently an Assistant Professor of Psychology at Emory University. Her research concerns spatial and numerical cognition. She is particularly interested in how young children specify location, embodied representations of space, sex differences in spatial reasoning, and interactions between space and number.

TEENIE MATLOCK earned a Ph.D in Cognitive Psychology in 2001 from University of California Santa Cruz and did post-doctoral research at Stanford University. She is currently Founding Faculty and Assistant Professor of Cognitive Science at University of California Merced. Her research interests include lexical semantics, metaphor, and perception and action. Her research articles span psycholinguistics, cognitive linguistics, and human-computer interaction.

KELLY S. MIX received her Ph.D in developmental psychology from the University of Chicago in 1995. She co-authored the book *Quantitative Development in Infancy and Early Childhood* (Oxford University Press, 2002), and has published numerous articles on cognitive development. In 2002 she received the Boyd McCandless Award for early career achievement from the American Psychological Association (Div. 7). She is an Associate Professor of Educational Psychology at Michigan State University.

KIRSTEN O'HEARN received her Ph.D in Experimental Psychology from the University of Pittsburgh in 2002, with a focus on cognitive development in infancy. After a NICHD-funded postdoctoral fellowship in the Department of Cognitive Science at Johns Hopkins University, she returned to Pittsburgh and is now an Assistant Professor of Psychiatry at the University of Pittsburgh School of Medicine. She studies visual processing in developmental disorders, examining how object representation and visuospatial attention may differ over development in people with Williams syndrome and autism.

MICHAEL RAMSCAR received his Ph.D in Artificial Intelligence and Cognitive Science from the University of Edinburgh in 1999. He has been on the faculty at Stanford since 2002. In his research he seeks to understand how our everyday notions of concepts, reasoning, and language arise out of the mechanisms of learning and memory as their architecture develops in childhood.

DANIEL C. RICHARDSON studied psychology and philosophy at Magdalen College, Oxford as an undergraduate, and received his Ph.D in psychology from Cornell University in 2003. After a postdoctoral position at Stanford, he was an assistant professor at University of California Santa Cruz, and then a lecturer at Reading University in the UK. Currently, he is a lecturer at University College London. His research studies the speech, gaze, and movements of participants in order to investigate how cognition is interwoven with perception and the social world.

LARISSA K. SAMUELSON is an Assistant Professor in the Department of Psychology at the University of Iowa. She is also affiliated with the Iowa Center for Developmental and Learning Sciences. She received her doctorate in Psychology and Cognitive Science from Indiana University in 2000. Her research interests include word learning, category development, dynamic systems theory, and dynamic field and connectionist models of development. A current area of particular focus is the development of word learning biases and the role of stimuli, the current task context, and children's prior developmental and learning history in emergent novel noun generalization behaviors.

LINDA B. SMITH is Chancellor's Professor of Psychology at Indiana University. She earned her Ph.D in developmental psychology from the University of Pennsylvania in 1977. She has published over 100 articles and books on cognitive and linguistic development, including *A Dynamic Systems Approach to Development* (MIT Press, 1993) and *A Dynamic Systems Approach to Cognition and Action* (MIT Press, 1994). In 2007 she was elected to the American Academy of Arts and Sciences.

JOHN P. SPENCER is an Associate Professor of Psychology at the University of Iowa and the founding Co-Director of the Iowa Center for Developmental and Learning Sciences. He received a Sc.B with Honors from Brown University in 1991 and a Ph.D in Experimental Psychology from Indiana University in 1998. He is the recipient of the Irving J. Saltzman and the J. R. Kantor Graduate Awards from Indiana University. In 2003 he received the Early Research Contributions Award from the Society for Research in Child Development, and in 2006 he received the Robert L. Fantz Memorial Award from the American Psychological Foundation. His research examines the development of visuo-spatial cognition, spatial language, working memory, and attention, with an emphasis on dynamical systems and neural network models of cognition and action. He has had continuous funding from the National Institutes of Health and the National Science Foundation since 2001, and has been a fellow of the American Psychological Association since 2007.

MICHAEL J. SPIVEY earned a BA in Psychology from University of California Santa Cruz in 1991, and then a Ph.D in Brain and Cognitive Sciences from the University of Rochester in 1996, after which he was a member of faculty at Cornell University for 12 years. He is currently Professor of Cognitive Science at University of California, Merced. His research focuses on the interaction between language and vision, using the methods of eye tracking, reach tracking, and dynamic neural network simulations. This work is detailed in his 2007 book, *The Continuity of Mind*.

MARINA VASILYEVA is an associate professor of Applied Developmental Psychology at the Lynch School of Education, Boston College. She received her Ph.D. in Psychology from the University of Chicago in 2001. Her research interests encompass language acquisition and the development of spatial skills. In both areas, she is interested in understanding the sources of individual differences, focusing in particular on the role of learning environments in explaining variability in cognitive development.

CHEN YU received his Ph.D in Computer Science from the University of Rochester in 2004. He is an assistant professor in the Psychological and Brain Sciences Department at Indiana University. He is also a faculty member in the Cognitive Science Program and an adjunct member in the Computer Science Department. His research interests are interdisciplinary, ranging from human development and learning to machine intelligence and learning. He has received the Marr Prize at the 2003 Annual Meeting of the Cognitive Science Society, and the distinguished early career contribution award from the International Society of Infant Studies in 2008.

CARLOS A. ZEDNIK received a BA from Cornell University in Computer Science and Philosophy and an MA in Philosophy of Mind from the University of Warwick, and is currently a Ph.D candidate in cognitive science at Indiana University, Bloomington. His primary interests are the perceptual foundations of language and mathematics, and the philosophical foundations of the dynamical systems approach to cognition.

# Abbreviations

| | |
|---|---|
| AVS | Attentional Vector Sum model |
| CA | Category Adjustment |
| CCD | charge-coupled device (camera) |
| DFT | Dynamic Field Theory |
| fMRI | functional magnetic resonance imaging |
| HMM | Hidden Markov Model |
| MA | mental age |
| MOT | multiple object tracking task |
| PF | perceptual field |
| PSS | Perceptual symbol system |
| SES | socioeconomic status |
| SWM | working memory field |
| TOM | theory of mind |
| VOT | Voice onset time |
| VR | Virtual reality |
| WS | Williams syndrome |

# Section I
# Thinking Through Space

The chapters in this section converge on a common theme. Because cognition happens during movement through space, space constitutes a major format of abstract thought. This claim goes beyond simply saying that people retain spatial information as part of their memories. It argues, instead, that we actively use space as a medium for interpretation, planning, and recall, whether or not the task at hand is inherently spatial (e.g. map reading).

For proponents of embodied cognition, there is nothing controversial here. As Clark points out, embodiment is what happens in real time and space, so it is natural to think that time and space would be incorporated into our mental representations in meaningful ways. These ideas are, therefore, entirely consistent with theories of perceptually grounded cognition (Barsalou 1999; Gibbs 2006; Glenberg 1997; Lakoff & Johnson 1980). However, it is worth taking a step back to recognize that acceptance of such claims is a relatively recent phenomenon. It was not so long ago that human thought was generally described in terms of logical propositions. Sensory or proprioceptive memory was considered incidental to learning, not central to it. Indeed, most developmentalists have viewed reliance on concrete experience as an indication of cognitive immaturity (e.g. Piaget). Such a view is practically antithetical with the idea that all thought, even the 'abstract' thinking of adults, derives from bodily experience. Thus, these chapters contribute to a theoretical movement that is new and quite distinctive.

One of the main contributions of the present work is to review the growing body of empirical support for embodiment. Spivey, Richardson, and Zednik describe a series of eye-tracking experiments that reveal the unexpected ways adults recruit physical space to remember verbal information. They also report several studies showing that people understand seemingly abstract verbs, such as 'respect' or 'succeed', in terms of vertical or horizontal spatial relations. Ramscar, Boroditsky, and Matlock present numerous experiments demonstrating that when people think about movement, as they might when taking a trip on an airplane, it changes the way they interpret ambiguous statements about time. Clark cites several studies, such as Beach's (1988) experiment with expert bartenders and distinctively shaped glasses, revealing the extent to which complex tasks require environmental scaffolding. Although most of the work summarized in these chapters has appeared separately elsewhere, the main findings are integrated here in a way that highlight their theoretical significance with crystal clarity. They provide powerful and convincing support for key predictions in the embodiment literature. In particular, the chapters by Spivey et al. and by Ramscar et al. are empirical *tours de force* that leave little doubt about the role of space in higher cognition.

In keeping with the theme of this volume, the chapters in this section also consider the way space and language might relate to one another within an embodiment framework. Clark sees language and space as performing similar functions: both serve to reduce complexity in the environment. After demonstrating that people routinely use spatial arrangements to scaffold challenging tasks, he points out that words can be used to indicate groupings without requiring objects to be gathered in space. He also argues that language and space are so tightly related in the service of this function that effects of space on language, and language on space, are nearly inevitable. For example, talking about fall foliage highlights certain features of the environment (e.g. trees, red) and thereby alters our perception—language parses the continuous flow of experience into meaningful units.

The remaining chapters focus on the use of space to ground language—particularly terms that are arguably less concrete by nature. As noted above, Spivey et al. demonstrate that abstract verbs are understood in terms of spatial relations. They also show that when adults recall verbal information, they look toward locations in their physical space that either contained visual information when the words were spoken or were used by the listeners to interpret language as it was presented. Thus, there is more to remembering words than processing the words themselves. Instead, people seem to actively recruit space to interpret and remember verbal symbols (words).

Similarly, Ramscar and colleagues focus on the way space might ground words for time. Unlike concepts with clear sensory correlates, such as texture or sweetness, time has an intangible quality. In fact, one could argue that time is so abstract that, until people can impose conventional units of time (learned via language) on their perceptual reality, they literally fail to experience it. At the least, we know from developmental studies that it takes children a long time to acquire the language of time, suggesting that the referents for these terms are not obvious (e.g. Friedman 2003). Ramscar et al.'s research shows that adults not only use motion to ground concepts of time, but do so in a fluid way that is sensitive to recent bodily experience. These authors speculate that the tight linkage between time words and space concepts arises from a common ancestry: notions of time and space both emerge from the concrete experience of movement. Thus, like Clark, they predict multiple intrusions and constraints of one type of thought on the other. In essence, they see these as different mental construals of the same physical experience.

Mix focuses on the language for another abstract domain: mathematics. Like time, one could argue that number and mathematical relations are difficult to perceive directly. When one enters a room, for example, many possible mathematical relations could be considered. There is no objective 'three', but there may be three tables or three plates, if the perceiver chooses to group and then enumerate them. There are a multitude of potential equivalencies, ordinal relations, combinations, or subtractions, but these also are psychologically imposed. Thus, the referents for mathematical language are not obvious. But rather than asking whether people spontaneously recruit space to ground these concepts, Mix considers whether children can be taught to do so by providing carefully designed spatial models. In essence, she asks whether the natural human tendency to think about language in terms of space can be harnessed and exploited by teachers.

Although the four chapters view space and language from slightly different angles, they all assume that the two are tightly coupled, even when people are not talking about space, and that this tight coupling comes from hearing words while experiencing bodily movements in space. Of course, rather than explaining how these linkages come to be, this assumption shifts the explanatory burden to developmental psychology and leaves many unanswered questions in its wake.

One of these is what conditions promote this coupling. In other words, what happens early in language development to encourage the conflation of space and words? Although the chapters in this section do not address this question directly, they hint that spatial grounding happens because there is no way to avoid it. Space is always with us—there is nothing we can experience that lacks a spatial dimension. Thus, words have to be grounded in space, because they refer to experiences that occur, by necessity, in space. The problem is that space is not the only omnipresent stream of information. People constantly experience a multitude of other percepts—color, texture, sound, intensity. How much of this information is retained in memory? What determines which pieces stay and which will go? And if space is truly special, what makes it so?

One possibility is that space really isn't all that special. It simply rises to the forefront of the present work because the authors designed their studies to tap spatial representation. For example, Spivey et al. chose to test verticality as the perceptual dimension that might ground verbs, but perhaps we would find similar effects for non-spatial perceptual dimensions as well. Maybe some verbs are hot and others are cold.

Alternatively, space may be genuinely unique in ways that make it a dominant medium of thought. But if it's not just because space is always there, why is it? Perceptual learning theorists contend that selective attention is the gatekeeper that raises or limits the salience of certain streams of input (e.g. Barsalou 1999; Goldstone 1998; Smith, Jones, & Landau 1996). People learn to fit their perceptions to certain tasks and use attention weighting to achieve this adaptation. This process involves learning to see distinctions as well as units—essentially drawing psychological boundaries around chunks of information that derive meaning from their utility. This framework suggests several routes by which space may become central to abstract thought and language.

One possibility is that first words are initially grounded in space by association. Infants and toddlers spend a lot of their time manipulating objects and gaining control over their own movements. In that sea of sensorimotor experience, they also hear words. The words that stick first are the ones that label objects and actions to which children are already attending (Baldwin 1991; Smith 2000; Yu & Ballard, Chapter 10 below). So, perhaps language and space become intertwined because children's first associations between symbol and meaning are forged in a multimodal soup—one that is heavily laden with spatial relations by virtue of the kinds of things toddlers naturally do. This may bias children to interpret words in terms of space even when the words no longer directly refer to spatial experience. If so, one should find that early vocabularies contain many words with an explicit spatial component. We should also see a progression from words about space to spatial metaphors—precisely what some symbol grounding theories would predict (e.g. Barsalou 1999; Lakoff & Nunez 2000).

Another possibility is that people learn to retain space and ignore other information because they are sensitive to the speed-accuracy trade-offs involved in using various scaffolds. If space consistently leads to faster, more accurate performance, it will be retained, whereas less useful dimensions (e.g. texture?) may be ignored. If attention to space leads to better performance in a wider range of tasks, or in tasks that are encountered more frequently, it may be retained in all kinds of representations simply because its salience is heightened a lot of the time. This suggests that space as a tool for thought may not be immediately 'transparent' in Clark's sense. Instead, it may take quite a lot of experience before people (especially children) zero in on space or become efficient at implementing it. And the use of words to control space likely develops even later—partly because children master movement through space earlier than they master language, but also because this interpretation implies a certain level of strategy (i.e. the idea that spatial groupings can facilitate processing seems to be a logical prerequisite to the idea that words can manipulate space in helpful ways).

A third possibility is that spatial information may simply be easier to process than other kinds of information. Rather than learning that spatial information is useful, and hence increasing one's attention to it, perhaps human beings naturally focus on space because it takes less effort to process than other percepts from the outset. In other words, our brains may be innately attuned to space more than they are to other percepts. This explanation brings us back to the 'spatial thought is inevitable' stance. It comes closest to the idea that abstract concepts are connected to space at a deep, unconscious level—literally the product of neural juxtaposition (Barsalou 1999; Spivey et al., Chapter 2 below). If this is the case, then we might expect to see an overreliance on spatial information early in development—children focusing on space when it is not necessary, to the exclusion of less accessible, but possibly more relevant information.

This analysis brings up several additional developmental issues beyond the question of how language and space become connected. One is whether the use of spatial metaphors is effortful and how this changes over development. Construing space as a tool (as Clark and Mix have done) implies at least the possibility that space might be recruited strategically. Certainly, the experts who design math manipulatives are purposefully recruiting space. But if thought is inherently spatial—if space is so integral to thought that the two are inseparable—is it possible to manipulate it purposefully? Do participants in Spivey et al.'s experiments know they are looking to a particular location to jog their memories? Do children become more able to implement spatial tools as they gain meta-awareness of their own cognition? Or do they become less adept as the tools they have acquired fade into automaticity? And if these behaviors are automatic and subconscious, is it realistic for educators to think they can harness it?

A related issue is whether children discover spatial scaffolding on their own, or acquire it from adults. Certainly, children witness adults arranging objects spatially to perform various tasks. Does this encourage them to do the same? When teachers use concrete models to illustrate difficult concepts, are they simply capitalizing on the human tendency to think in spatial terms, or are they explicitly teaching children to use spatial metaphors? Is that instruction necessary?

Finally, we might wonder about the potential mismatches between one person's spatial groundings and another's. If words are spatially grounded, and the same words mean roughly the same thing to different people, then we must assume that the underlying spatial metaphors are roughly the same. How does this consistency develop? How much variability is permissible before communication breaks down? This issue has major implications for teaching with spatial models, because models can only be effective if they activate relevant, embodied representations. Typically, math manipulatives have many details stripped away, but does this make them more difficult to link up with other experiences? How can teachers know whether the model they have created is commensurable with the prior knowledge of their students?

In summary, the four chapters in this section go a long way toward establishing that adults recruit space to ground language and identifying the implications of that realization. However, the growth edge for this line of research—where it can move toward a more satisfying level of explanation—resides in the basic developmental questions it raises. By understanding how spatial grounding develops, we will know more about the forces that make it happen and the resulting shape these representations are likely to take.

# 1

## Minds in Space

ANDY CLARK

In what ways might real spatiality impact cognition? One way is by providing a resource for the intelligent offloading of computational work. Space is a prime resource for *cognitive niche construction*. In this chapter I examine some of the many ways this might work, and then pursue some structural analogies between what David Kirsh (1995) has called 'the intelligent use of space' and the intelligent use of language. By this I mean the use of language not just as a communicative instrument, but as a means of altering and transforming problem spaces (see Clark 1998). Spatial and linguistic manipulations can each serve to reduce the descriptive complexity of the environment. I ask whether this parallel is significant, and whether one function of talk itself is to provide a kind of vicarious restructuring of space (a low-tech version of augmented reality).

### 1.1 Space

Space and language are usefully conceived as interacting and mutually supporting forces in the process of cognitive niche construction: the process of actively structuring a world in which to think. But before this possibility can come into view, we need to ask what we mean by space and by 'real spatiality' anyway. I suggest that real space *just is* wherever perception and embodied action can occur. Spatiality and embodiment, on this account, always go hand in hand. Such a view is convincingly developed by Dourish (2001), who begins by offering the following formulation for the notion of embodiment itself:

Embodiment 1: Embodiment means possessing and acting through a physical manifestation in the world. (Dourish 2001: 100)

Unpacking this in turn, we quickly read:

Embodiment 2: Embodied phenomena are those that by their very nature occur in real time and real space. (p. 101)

To see what is meant by this, we are asked to consider the contrast between what Dourish calls 'inhabited interaction' and 'disconnected control'. Since this bears rather directly upon the notion of real spatiality that I shall offer shortly, it is worth reviewing the passage in full:

Even in an immersive virtual-reality environment, users are disconnected observers of a world they do not inhabit directly. They peer out at it, figure out what's going on, decide on some course of action, and enact it through the narrow interface of the keyboard or the data-glove, carefully monitoring the result to see if it turns out the way they expected. Our experience in the everyday world is not of that sort. There is no homunculus sitting inside our heads, staring out at the world through our eyes, enacting some plan of action by manipulating our hands, and checking carefully to make sure we don't overshoot when reaching for the coffee cup. We inhabit our bodies and they in turn inhabit the world, with seamless connections back and forth. (p. 102)

I do not believe that immersive virtual reality (VR) is by its very nature disconnected in this sense. Rather, it is just one more domain in which a skilled agent may act and perceive. But skill matters, and most of us are as yet unskilled in such situations. Moreover (and this is probably closer to Dourish's own concerns), the modes of interaction supported by current technologies can seem limited and clumsy, and this turns the user experience into that of a kind of alert game-player rather than that of an agent genuinely located inside the virtual world.

It is worth noticing, however, that to the young human infant, the physical body itself may often share some of this problematic character. The infant, like the VR-exploring adult, must learn how to use initially unresponsive hands, arms, and legs to obtain its goals. With time and practice, this all changes, and the problem space is now not that of the body so much as the wider world that the body makes available as an arena for action. At this moment, the body has become what some philosophers, influenced by Heidegger (1961[1927]), call 'transparent equipment'. This is equipment (the classic example is the hammer in the hands of the skilled carpenter) that is not the focus of attention in use. Instead, the user 'sees through' the equipment to the task in hand. When you sign your name, the pen is not normally your focus (unless it is out of ink, etc.). The pen, in use, is no more the focus of your attention than is the hand that grips it. Both are transparent equipment.

What really matters for my purposes, though, is one very distinctive feature of transparent equipment. Transparent equipment presents the world to the user not just as a problem-space (though it is clearly that) but also as a resource. In this way the world, encountered via transparent equipment, is a place in which we can act fluently in ways that simplify or transform the problems that we want to solve. According to this diagnosis, what makes us feel like visitors to VR-space (rather than inhabitants) is that our lack of skill typically (though by no means necessarily) forces us to act effortfully and to reason *about* the space, rather than to act easily, and to reason *using* (instead one could say *in* the space). This 'intelligent use of space' (Kirsh 1995) is the topic of the next section.

Using these ideas, we can now motivate a unified and liberal account of embodiment and of real spatiality. By a 'unified' account, I mean one in which the definition of embodiment makes essential reference to that of space, and vice versa, so that the two are co-defining (like the concepts of buying and selling). By a 'liberal' account, I mean one that does not simply assume that standard human bodies and standard physical three-space are essential to either 'real space' or 'genuine embodiment'. The space our bodies inhabit is defined by the way it supports fluent action, and what this means (at least in part) is that it is defined by the way it presents the world as a resource for reasoning rather than simply a domain to be reasoned about.

## 1.2 Space as a resource for reasoning

Human beings are remarkably adept at the construction and reconstruction of their own cognitive niches. They are adept at altering the world so as to make it a better place in which to think. Cognitive niche construction, thus understood, is the process by which human inventions and interventions sculpt the social, symbolic, and physical environment in ways that simplify or productively transform our abilities to think, reason, and problem-solve.

The idea of humans as cognitive niche constructors is familiar within cognitive science. Richard Gregory (1981) spoke of 'cognition amplifiers', Don Norman (1993) of 'things that make us smart', Kirsh & Maglio (1994) of 'epistemic actions', Daniel Dennett (1996) of 'tools for thought': the list could be continued. One of my own favorite examples (from Clark 1997) concerns the abilities of the expert bartender. Faced with multiple drink orders in a noisy and crowded environment, the expert mixes and dispenses drinks with amazing skill and accuracy. But what is the basis of this expert performance? Does it all stem from finely tuned memory and motor skills? In controlled psychological experiments comparing novice and expert bartenders (Beach 1988, cited in Kirlik 1998: 707), it becomes clear that expert skill involves a delicate interplay between internal and environmental factors. The experts select and array *distinctively shaped glasses* at the time of ordering. They then use these persistent cues so as to help recall and sequence the specific orders. Expert performance thus plummets in tests involving uniform glassware, whereas novice performances are unaffected by any such manipulations. The expert has learned to sculpt and exploit the working environment in ways that transform and simplify the task that confronts the biological brain.

This is a clear case of 'epistemic engineering': the bartender, by creating persisting spatially arrayed stand-ins for the drinks orders actively structures the local environment so as to press more utility from basic modes of visually cued action and recall. In this way, the exploitation of the physical situation allows relatively lightweight cognitive strategies to reap large rewards. Above all, it is a case in which we trade active local spatial reorganization against short-term memory.

This is by no means an isolated case. A vast amount of human cognitive niche construction involves the active exploitation of space. David Kirsh, in his classic treatment 'The intelligent use of space' (1995), divides these uses into three broad (and overlapping) categories.

The first is 'spatial arrangements that simplify choice', such as laying out cooking ingredients in the order you will need them, or putting your shopping in one bag and mine in another. The second is 'spatial arrangements that simplify perception', such as putting the washed mushrooms on the right of the chopping board and the unwashed ones on the left, or the color green-dominated jigsaw puzzle pieces in one pile and the red-dominated ones in another. The third is 'spatial dynamics that simplify internal computation', such as repeatedly reordering the Scrabble pieces so as to prompt better recall of candidate words, or the use of instruments such as slide rules, which transform arithmetical operations into perceptual alignment activities.

Kirsh's detailed analysis is concerned solely with the adult's expert use of space as a problem-solving resource. But it is worth asking how and when children begin to use active spatial reorganization in this kind of way. Is this something that we, as humans, are just naturally disposed to do, or is it something we must learn? A robot agent, though fully able to act on its world, will not *ipso facto* know to use space as a resource for this kind of cognitive niche construction! Indeed, it seems to me that no other animal on this planet is as adept as we are at the intelligent use of space: no other animal uses space as an open-ended cognitive resource, developing spatial offloadings for new problems on a day-by-day basis. A good question is thus: Just what do you need to know (and to know how to do) to use space as an open-ended cognitive resource?

I do not have answers to these questions, but I do have one very speculative suggestion, which I will offer only in the spirit of the brainstorming recommended by the organizers. It is noteworthy, it seems to me, that the majority of the spatial arrangement ploys work, as Kirsh himself notes at the end of his long treatment, by *reducing the descriptive complexity of the environment.*

Space is used as a resource for grouping items into equivalence classes for some purpose (*washed* mushrooms, *red* jigsaw pieces, *my* shopping, and so on). It is intuitive that once descriptive complexity is reduced, processes of selective attention, and of action control, can operate on elements of a scene that were previously too 'unmarked' to define such operations over. The (very) speculative idea that I want to float is that humans may have an innate drive to reduce the descriptive complexity of their worlds, and that such a drive (vague and woolly though that idea is) might also be part of what leads us to develop human-like language. For human language is itself notable both for its open-ended expressive power and for its ability to reduce the descriptive complexity of the environment. Reduction of descriptive complexity, however achieved, makes new groupings available for thought and action. In this way, the intelligent use of space and the intelligent use of language may form a mutually reinforcing pair, pursuing a common cognitive agenda.

## 1.3 Space and language

The cognitive functions of space and language are strikingly similar. Each is a resource for reducing descriptive complexity. Space works by means of physical groupings that channel perception and action towards functional or appearance-based equivalence classes. Language works by providing labels that pick out all and only the items belonging to equivalence classes (the red cups, the green balls, etc.). Both physical and linguistic groupings allow selective attention to dwell on all and only the items belonging to the class. It is fairly obvious, moreover, that the two work in fairly close cooperation. Spatial groupings are used in teaching children the meanings of words, and words are used to control activities of spatial grouping.

Once word learning is under way, language begins to function as a kind of augmented reality trick by means of which we cheaply project new groupings and structures onto a perceived scene. By 'cheaply' I mean: first, without the physical effort of putting the linguistically grouped items into piles (saying 'the yellow flowers' is thus like grouping all the yellow flowers in one place and then, for good measure, adding a distinctive flag to that pile); and second, without effective commitment to a single persisting classification. It is cheap and easy to first say 'Look at all the yellow flowers on her hat' and then 'Look at all the different-colored flowers on all the hats', whereas real spatial groupings (say, of dolls, hats, and flowers) would require several steps of physical reorganization.

Linguistic labels, on this view, are tools for grouping, and in this sense act much like real spatial reorganization. But in addition (and unlike physical groupings), they effectively add new items (the overlaid labels themselves) to the scene. Language thus acts as a source of *additional* cues in a matrix of multi-cued problem-solving. This adds a very special layer of complexity to language-mediated cognitive niche construction.

A simple demonstration of this added complexity is found in Thompson, Oden, & Boysen (1997). In a striking experiment, language-naïve chimpanzees (Pan troglodytes) were trained to associate a simple plastic token (such as a red triangle) with any pair of identical objects (two shoes, say) and a differently shaped plastic token with any pair of different objects (a cup and a shoe, or a banana and a rattle). The token-trained chimps were subsequently able—without the continued use of the plastic tokens—to solve a more complex, abstract problem that baffled non-token-trained chimps. The more abstract problem (which even we sometimes find initially difficult!) was to categorize *pairs-of-pairs* of objects in terms of *higher-order* sameness or difference. Thus the appropriate judgment for the pair-of-pairs 'shoe/shoe and banana/shoe' is 'different' because the *relations* exhibited within each pair are different. In shoe/shoe the (lower order) relation is 'sameness'. In banana/shoe it is 'difference'. Hence the higher-order relation—the relation *between* the relations—is difference. By contrast, the two pairs 'banana/banana and cup/cup' exhibit the higher-order relation 'sameness', since the lower-level relation (sameness) is the same in each case.

To recap, the chimps whose learning environments included plastic tokens for sameness and difference were able to solve a version of this rather slippery problem. Of the chimps not so trained, not a single one ever learned to solve the problem. The high-level, intuitively more *abstract,* domain of relations-between-relations is effectively invisible to their minds. How, then, does the token-training help the chimps whose early designer environments included plastic tokens and token-use training?

Thompson et al. (1997) suggest that the chimp's brains come to associate the 'sameness' judgements with an inner image or trace of the external token itself. To be concrete, imagine the token was a red plastic triangle and that when they see two items that are the same they now activate an inner image of the red plastic triangle. And imagine that they associate judgements of difference with another image or trace (an image of a yellow plastic square, say). Such associations reduce the tricky higher-level problems to lower-order ones defined not over the world but over the inner images of the plastic tokens. To see that 'banana/shoe' and 'cup/apple' is an instance of higher-order sameness, all the brain now needs to do is recognize that two green triangles exhibit the lower-order relation sameness. The learning made possible through the initial loop into the world of stable, perceptible plastic tokens has allowed the brain to build circuits that reduce the higher-order problem to a lower-order one of a kind their brains are already capable of solving.

Notice, finally, that all that is really needed to generate this effect is the association of the lower-order concepts (sameness and difference) with stable, perceptible items. What, then, is the spoken language we all encounter as infants if not a rich and varied repository of such stable, repeatable auditory items? The human capacity for advanced, abstract reason surely owes an enormous amount to the way these words and labels act as a new domain of simple objects on which to target our more basic cognitive abilities. *Experience* with external tags and labels is what enables the brain itself—by *representing* those tags and labels—to solve problems whose level of complexity and abstraction would otherwise leave us baffled.

Learning a set of tags and labels (which we all do when we learn a language) is a key means of reducing the descriptive complexity of the environment by rendering certain features of our world concrete and salient. Just like the simple trick of spatial grouping, it allows us to target our thoughts (and learning algorithms) in new ways. But in addition, the labels themselves come to constitute a whole new domain of basic objects. This new domain compresses what were previously complex and unruly sensory patterns into simple objects. These simple objects can then be attended to in ways that quickly reveal further (otherwise hidden) patterns, as in the case of relations-between-relations. And of course the whole process is deeply iterative: we coin new words and labels to concretize regularities that we could only originally conceptualize thanks to a backdrop of other words and labels.

In sum, words and labels help make relations we can perceptually detect into objects, allowing us to spot patterns that would otherwise elude us. We can think of this as a kind of augmented reality device that projects new perceptible structures onto the scene, allowing us to reap some of the benefits of physical grouping and marking without actually intervening in the environment. Moreover, the labels, when physically realized (as plastic tokens or word inscriptions) are themselves genuine objects apt *both for perception and spatial reorganization*. There is the hint here of a synergy so potent that it may form a large part of the explanation of our distinctively human form of intelligence. Perhaps, then, language and the actual use of space (for grouping and regrouping during learning and problem-solving) form a unified cognitive resource whose function is the reduction of descriptive complexity via the dilation, compression, and marking of patterns in complex sensory arrays.

The power and scope of human reason owes much, I suspect, to the action of this unified (spatio-linguistic) resource. But our understanding of it is undeveloped. We look at the parts (space, language) but not at the inter-animated whole. Thinking of language itself as a kind of cheap medium for the vicarious re-structuring of perceptual space may be a way to bring the elements into a common framework for study and model-building.

## 1.4  The evolution of space

I would like to touch on one final topic before ending, and that is the nature of space itself. Space, as I have defined it, is an arena for embodied action (and embodied action is essentially action in space, and time). But nothing in this definition commits us to any specific form of embodiment, or of spatiality. What matters is to be able to interact with a stable and exploitable resource upon which to offload computational work. As new technologies increasingly blur the boundaries between the physical and the digital worlds, the way space and body are themselves encountered during development may alter and evolve. We may become adept at controlling many kinds of body, and exploiting many kinds of space. Cognitive niche construction can very well occur in hybrid physical/informational worlds: imagine organizing information for visual retrieval in a virtual reality environment in which the laws of standard three-space do not apply, and infinite (and perhaps recursive) stacking of objects within objects is possible! Inhabited interactions with such a world are, I believe (see Clark 2003) entirely possible. Certainly, as the worlds of digital media and everyday objects begin to blur and coalesce, we will develop new ways of acting and intervening in a hybrid 'digital-physical' space. This hybrid space will be the very space we count (by my definitions) as being embodied within. New opportunities will exist to use this combined space as a cognitive resource—for example, by using genuine augmented reality overlays as well as real spatial organization and linguistic labeling. Understanding cognitive development in a hybrid (physical-digital) world may thus be a major task for the very near future. Understanding the complex interplay of space, language, and embodied action in the standard case is probably essential if we are to make the most (and avoid the worst) of these new opportunities.

## 1.5  Conclusions

In this speculative and preliminary treatment, I have tried to put a little flesh on the idea of space and language forming a unified cognitive resource. Spatiality and language, I suggested, may be mutually reinforcing manipulanda (to borrow Dan Dennett's useful phrase): cognitive tools that interact in complex ways so as to progressively reduce the descriptive complexity of the problem-solving environment.

Open questions for future research include: How do we learn to make intelligent use of space? How do the intelligent use of space and of language interact? Might both be rooted in an innate drive to reduce descriptive complexity? Does it help to consider language as a means of vicariously restructuring the perceptual array for cognitive ends, and space as a resource for physically achieving the same goal? What happens when linguistic structure is *itself* encountered as a spatial array, as words on a page, or labels in a picture book? Are these good questions to ask? It seems too early to say. But understanding the role of real space in the construction and operation of the mind is essential if we are to take development, action, and material structure seriously, as the essence of real-world cognition rather than its shallow surface echo.

2

# Language Is Spatial, Not Special: On the Demise of the Symbolic Approximation Hypothesis

MICHAEL J. SPIVEY, DANIEL C. RICHARDSON, AND
CARLOS A. ZEDNIK

In this chapter, we argue that cognitive science has made as much progress as possible with theories of discrete amodal symbolic computation that too coarsely approximate the neural processes underlying cognition. We describe a collection of studies indicating that internal cognitive processes are often constructed in, and of, analog spatial formats of representation, not unlike the topographic maps that populate so much of mammalian cortex. Language comprehension, verbal recall, and visual imagery all appear to recruit particular spatial locations as markers for organizing, and even externalizing, perceptual simulations of objects and events. Moreover, not only do linguistic representations behave as if they are *located* in positions within a two-dimensional space, but they also appear to *subtend* regions of that space (e.g. perhaps elongated horizontally or vertically). This infusion of spatial formats of representation for linguistically delivered information is particularly prominent in the analyses of cognitive linguistics, where linguistic entities and structures are treated not as static logical symbols that are independent of perception and action but instead as spatially dynamical processes that are grounded in perception and action. Some of the predictions of this framework have recently been verified in norming studies and in experiments showing online effects of linguistic image schemas on visual perception and visual memory. In all, this collection of findings points to an unconventional view of language in which, far from being a specialized modular mental faculty performing computations on discrete logical symbols, linguistic ability is an emergent property that opportunistically draws from the existing topographic representational formats of perceptual and motor processes.

As originally conceived, the behavioral mimicry arose from the underlying mimicry between biological neurons and switchlike elements, and on a continuity assumption, or robustness hypothesis, that populations of comparable elements arrayed comparably would behave in comparable ways. This kind of plausibility has been entirely lost in the progression from neural net through finite automaton through Turing machine, in which comparability devolves entirely on behaviors themselves, rather than on the way the behaviors are generated. In mimetic terms, we now have actors (e.g. Turing machines) imitating actors (automata) imitating other actors (neural nets) imitating brains. What looks at each step like a gain in generality (i.e. more capable actors) progressively severs every link of plausibility…(Rosen 2000: 156)

## 2.1  The Symbolic Approximation Hypothesis

In the early years of cognitive science, the few pioneers who were concerning themselves with neurophysiology staked their careers on the assumption that populations of spiking neurons would behave more or less the same as populations of digital bits (e.g. Barlow 1972; McCulloch 1965; Von Neumann 1958; Wickelgren 1977; see also Lettvin 1995; Rose 1996). This assumption is what Rosen (2000) refers to in the quotation above: 'populations of comparable elements arrayed comparably would behave in comparable ways.' However, a great deal more has been learned in the past few decades about how populations of neurons work (e.g. Georgopoulos, Kalaska, Caminiti, & Massey 1982; Pouget, Dayan, & Zemel 2000; Sparks, Holland, & Guthrie 1976; Tanaka 1997; Young & Yamane 1992), and it is nothing at all like the instantaneous binary flip-flopping from one discrete state to another that characterizes the 'switchlike elements' of digital computers. The individual neurons that make up a population code do not appear to update their states in lockstep to the beat of a common clock (except perhaps under spatially and temporally limited circumstances: Engel, Koenig, Kreiter, Schillen, & Singer 1992; but cf. Tovee & Rolls 1992). Population codes spend a substantial amount of their time in *partially* coherent patterns of activity. And the brain's state is often dynamically traversing intermediate regions of the contiguous metric state space that contains its many semi-stable attractor basins. The distant and tenuous mimetic connection between symbolic computation and the brain is precisely what Rosen excoriates. In this chapter, we shall refer to this fragile link as the Symbolic Approximation Hypothesis. According to this hypothesis, mental activity is sufficiently approximated by models that use rule-based operations on logical symbols, despite the fact that empirical evidence suggests that neither the intensional *contents* nor the physical *vehicles* of mental representations are consistent with this approximation.[1]

There are two key properties of the representational contents instantiated by neural populations that separate them from the contents of symbolic representations: (1) continuity in time, and (2) continuity in space. Analogously, these same two properties are exhibited by the neural vehicles carrying representational contents. Page limits force us to restrict our discussion here to the property of continuity in space, and even then only to the representational content component. Continuity in time is dealt with elsewhere in two different ways, corresponding to the vehicle/content distinction: (a) the continuous temporal dynamics of the neural connectivity patterns, that constitute the vehicle of knowledge and intelligence, changing over developmental time (e.g. Elman, Bates, Johnson, Karmiloff Smith, Parisi, & Plunkett 1996; Spencer & Schöner 2003; Thelen & Smith 1994), and (b) the continuous temporal dynamics of neural activation patterns (i.e. representations) and behavior in real-time processing (e.g. Kelso 1994; Port & van Gelder 1995; Spivey 2007).

Continuity in space is dealt with in these two ways as well: (a) a *high-dimensional* state space in which the structure of the neural connections can be mathematically described as a contiguous attractor manifold. (e.g. Aleksander 1973; Lund & Burgess 1996; Edelman 1999; Elman 1991; Pasupathy & Connor 2002); and (b) a *two-dimensional* representation of space based on sensory surfaces, in which the intensional shape and layout of internal representations are roughly homologous to actual physical patterns of stimulation (e.g. Farah 1985; Kosslyn Thompson, Kim, & Alpert 1995; see also Barsalou 1999). The focus of the present chapter is on this latter type of spatial continuity: a very general notion of a two-dimensional spatially contiguous medium of representation. In some disciplines this is realized as 'topographic maps' for perception (e.g. Kohonen 1982; Swindale 2001; von der Malsburg 1973), in other disciplines as 'mental models' for cognition (Bower & Morrow 1990; Johnson-Laird 1983; 1998; Zwaan 1999), and still other disciplines as 'image schemas' for language (e.g. Gibbs 1996; Lakoff 1987; Langacker 1990; Talmy 1983).

---

[1] There is a related, but importantly different, perspective that might be termed the Symbolic *Abstraction* Hypothesis, in which discrete mental states are seen as supervening on neural hardware but not solely determined by, or reducible to, that hardware (e.g,. Dietrich & Markman 2003). According to that perspective, mental representations and their underlying neural vehicles are structurally independent, suggesting that characteristics of the vehicles have no necessary relation to characteristics of the representations themselves. However, the Symbolic Abstraction Hypothesis is usually espoused by cognitive scientists who are decidedly opposed to "concerning themselves with neurophysiology," and their simultaneous disavowal of Cartesian dualism while maintaining the independence of the mental and physical levels of explanation is becoming a difficult position to defend (cf. Kim 1998).

Let us assume that empirical evidence supporting the temporal continuity of representational vehicles and representational content is sound. Let us further assume that the claim of spatial continuity of representational vehicles is supported by the organization of neural structures underlying mental representations. Under such circumstances, it is clear that any evidence supporting the claim that representational *contents* are also spatially continuous would put a computational theory of cognition into a 'continuity sandwich', where that which is being computed, as well as that which is doing the computation, is continuous. In other words, if the vehicles as well as the contents of mental representations seem to be continuous in the temporal as well as the spatial sense, then any empirical justification for the Symbolic Approximation Hypothesis instantly evaporates—as it would paradoxically require continuous neural vehicles to represent continuous contents via *dis*continuous symbols and instructions. The 'topographic maps' view of mental representation that is presented here provides significant empirical evidence for the spatial continuity of representational content, thus ultimately suggesting that the Symbolic Approximation Hypothesis is in fact implausible.

It might be tempting to object to this argument by claiming that neither the continuous nature of physical vehicles nor the continuity of representational content is incompatible with symbolic approximation, since it seems nevertheless possible that discrete symbolic computational units are the *functionally relevant* units of a continuous neural substrate, and that they represent continuous intensional content. However, this objection fails on two counts. First and foremost, it is confusing the Symbolic Approximation Hypothesis with the Symbolic Abstraction Hypothesis mentioned earlier (see footnote 1): it fails to recognize the fact that, if the neurons are the vehicle and the action potentials of those neurons are the content, then an intimate connection between vehicle and content cannot help but exist. Secondly, this objection fails to consider the fact that discrete symbols often fail spectacularly at representing continuous information (cf. Bollt, Stanford, Lai, & Zyczkowski 2000). In other words, if the empirical argument for a continuous 'topographic maps' view of representational content is sound, then it is wrong to assume that this content is carried by discrete vehicles.

True symbol manipulation would require a kind of neural architecture that is very different from analog two-dimensional maps, such as individual 'grandmother cells', each devoted to a different concept (Lettvin 1995; see also Hummel 2001; Marcus 2001). So far, no neural areas or processes have been found in the primate brain that work in a way that would be genuinely amenable to pure rule-based computation of discrete logical symbolic representations (cf. Churchland & Sejnowski 1992; Rose 1996). Visual object and face recognition areas of the brain do appear to have cells that are substantially selective for particular objects or faces, but they also tend to be partially active for similar objects or similar faces (e.g. Gauthier & Logothetis 2000; Perrett, Oram, & Ashbridge 1998). Moreover, the continuous temporal dynamics involved in these cells gradually achieving, and contributing to, stable activity patterns (cf. Rolls & Tovee 1995) make it difficult for this process to be likened to symbolic computation.

Before much was known about neurophysiology and computational neuro-science, the Symbolic Approximation Hypothesis was a legitimate idealization, and probably a necessary one to get the ball rolling for cognitive science. However, in the past few decades, much has been learned about how real neural systems function (for reviews, see Churchland & Sejnowski 1992; O'Reilly & Munakata 2000), and it is substantially different from symbol manipulation. True Boolean symbol manipulation cannot take place in a spatially laid out arena of representation such as a topographic map where spatial proximity is a multi-valued parameter that constantly influences the informational content of the neural activity on the map. And when one surveys the neurophysiology literature, it becomes clear that topographic maps abound throughout the brain's sensory and motor cortices. As well as retinotopic, tonotopic, and somatotopic cell arrangements, there are topographic maps throughout the many higher-level cortices formerly known as 'associative', e.g. sensorimotor and polysensory cortices. Thus, it should not be too controversial to claim that much of perception and cognition is implemented in the two-dimensional spatial formats of representation that we know exist in the brain, without the use of binary symbolic representations that we have yet to witness.

## 2.2 Symbolic dynamics

Continuity in processing and representation runs counter to the system requirements for symbolic computation. Binary logical symbols must have qualitatively discontinuous boundaries that discretely delineate one from another. If symbols are allowed to exhibit partial continuous overlap with one another (in time and/or in space), then the rules being applied to them must become probabilistic (e.g. Rao, Olshausen, & Lewicki 2002) or fuzzy logical (e.g. Massaro 1998), which moves the theory substantially away from traditional notions of Turing Machines and at least partway toward analog, distributed, and dynamical accounts of mind.

Despite the difficulties involved in implementing genuine symbol manipulation in realistic neural systems, arguments for symbolic computation in the mind persist in the field of cognitive science (e.g. Dietrich & Markman 2003;

Hummel & Holyoak 2003; Marcus 2001; see also Erickson & Kruschke 2002; Sloman 1996). It is often implied that perception should be treated as the uninteresting stuff that uses analog representational formats in the early stages of information processing, and cognition is the interesting stage for which those analog representations must be converted into discrete symbols. Although the details of this conversion are typically glossed over in the psychological literature, a relatively young branch of mathematics may provide a promising framework in which this putative translation from analog-perceptual to discrete-conceptual can finally be rigorously worked out. In symbolic dynamics, a discretely delineated and internally contiguous region of state space, or phase space, can be assigned a symbol that is categorically different from the symbol assigned to a neighboring (and abutting) delineated region. As the continuous trajectory of the dynamical system's state moves into one region, the corresponding symbol is emitted, and when the trajectory then leaves that region and enters a different one, a new symbol is emitted (cf. Crutchfield 1994; Devaney 2003; Robinson 1995; for related treatments, see also Casey 1996; Cleeremans, Servan Schrieber, & McClelland 1989; Towell & Shavlik 1993). This framework entails two separate systems: (1) a continuous analog (perceptual) system that has relatively continuous internal dynamics of its own and is also influenced by external afferent input, and (2) a discrete binary (cognitive) system that receives the symbols emitted as a result of the state of the first system crossing one of its thresholds and entering a specific labeled region. Crucially, the symbolic system never receives any information about the particular state space coordinates of the continuous system. The emitted symbol is all it receives. Thus, it actually has a fair bit in common with the phenomena of categorical perception in particular (Harnad 1987), and it fits nicely with many traditional descriptions of the assumed distinction between perception and cognition in general (e.g. Robinson 1986; Roennberg 1990).

However, the devil, as always, is in the details. In symbolic dynamics, everything rests on the precisely accurate placement of the partition that separate one symbol's region from another symbol's region. The tiniest inaccuracies in partition placement can have catastrophic results. Even statistically robust methods for placement of partitions are often plagued by just enough noise to unavoidably introduce the occasional minor deviation in partition placement, which 'can lead to a severe misrepresentation of the dynamical system' (Bollt et al. 2000: 3524), resulting in symbol sequences that violate the system's own well-formedness constraints. Thus, while the field of symbolic dynamics is probably the one place where computational representationalists can use a mathematically explicit framework for exploring their psychological claims regarding the relationship between perception and cognition, that field may already be discovering problems that could unravel its promise for this particular use (cf. Dale & Spivey 2005).

Over and above the quantitative/statistical problems that arise with symbolic dynamics, there is an empirical concern to consider before applying symbolic dynamics to the supposed perception/cognition dichotomy. Although there is a great deal of concrete neuroanatomical and electrophysiological evidence for the distributed and dynamic patterns of representation in perceptual areas of the primate brain, there is no physical evidence for discrete and static symbolic representations in cognitive areas of the primate brain. Indeed, the areas of the primate brain that are thought to underlie cognition exhibit much the same kinds of distributed patterns of representation as in the areas thought to underlie perception and action (cf. Georgopoulos 1995). This lack of physical evidence casts some doubt on the 'cognitive bottleneck' idea that perception's job is to take the graded, uncertain, and temporarily ambiguous information in the sensory input and 'funnel' it into a finite set of discrete enumerable symbols and/or propositions that are posited with certainty and used by cognition in rule-based operations for logical inference.

## 2.3  Language is not special

This 'cognitive bottleneck' idea began developing much of its popularity around the time categorical speech perception was discovered (Liberman, Harris, Hoffman, & Griffith 1957). This particular phenomenon is perhaps most famous for popularizing the notion that 'speech is special' (Liberman 1982). For example, a stimulus continuum between the sound 'ba' and the sound 'pa', in which increments of voice onset time are used to construct intermediate sounds, is consistently perceived as two separate and monolithic categories of speech sounds, rather than as a gradual increase in voice onset time. Moreover, and perhaps more importantly, listeners are unable to discriminate between stimuli within a category, such as a 'ba' with 10 milliseconds VOT and a 'ba' with 20 milliseconds VOT. However, when the stimuli span the category boundary, such as a sound with 40 milliseconds VOT and 50 milliseconds VOT, discrimination is reliably above chance performance. Thus, the graded information within the category appears to be absent from the internal representation; all that is left is the category label (Liberman, Harris, Kinney, & Lane 1961; see also Dorman 1974; Molfese 1987; Simos, Diehl, Breier, Molis, Zouridakis, & Papanicolaou 1998; Steinschneider, Schroeder, Arezzo, & Vaughan 1995).

Perhaps the most famous breakdown for the popular interpretation of categorical speech perception as indicating that 'speech is special' was the finding that humans are not the only animals that exhibit categorical perception and discrimination of speech sound continua. Chinchilla and quail show the same effects (Kluender, Diehl, & Killeen 1987; Kuhl & Miller 1975). And since we probably would not want to use those results to conclude that chinchilla and quail have a special module devoted to language, perhaps we should not do the same for humans. Moreover, speech is not the only thing that exhibits these putatively categorical effects in perception. Colors appear to be perceived somewhat categorically as well (Bornstein & Korda 1984; 1985). And, finally, attractor networks are able to simulate categorical perception phenomena without the use of symbolic representations (Anderson, Silverstein, Ritz, & Jones 1977; see also Damper & Harnad 2000). Suddenly, speech no longer seems so 'special'.

Nonetheless, these categorical perception phenomena—although apparently not unique to speech or even to humans—are still commonly interpreted as exactly the kind of 'cognitive bottleneck' on which a discrete symbolic approach to cognition would naturally depend (Harnad 1987). However, there are a few hints in the categorical speech perception literature suggesting that the graded information in the stimulus is not completely discarded. Pisoni and Tash (1974) showed that when listeners are attempting to identify a sound that is on or near the boundary between these categories (between 30 and 50 milliseconds VOT), they take a longer time to make the identification, even though they systematically make the same identification almost every time. It is as though the two possible categories are partially represented simultaneously, like two mutually exclusive population codes that are each trying to achieve pattern completion and must compete against one another to do so. If they are nearly equal in their activation (or 'confidence'), they will compete for quite a while before one reaches a probability high enough to trigger its associated response, thus delaying the identification.

Another hint that graded information is actually still available in 'categorical' speech perception comes from work by Massaro (1987; 1998), on extending what is often called the McGurk effect (McGurk & MacDonald 1976; see also Munhall & Vatikiotis Bateson 1998). In the McGurk effect, the visual perception of a speaker's dynamic mouth shape has a powerful and immediate influence on the listener's perception of the phoneme being spoken. In Massaro's experimental framework, he presents to listeners a 'ba'/'da' continuum, where the place of articulation (what parts of the mouth constrict airflow during the sound) is varied in steps by digitally altering the speech waveform. That, by itself, tends to produce the standard categorical perception effect, as though the gradations in the stimuli are completely discarded by the perceiver. But Massaro couples this auditory 'ba'/'da' continuum with a computerized face, whose lips can be adjusted in steps along a *visual* 'ba'/'da' continuum (basically, by increasing the aperture between the lips). When these graded visual and auditory information sources are combined for perceiving the syllable, results are consistent with an algorithm in which the continuous biases in each information source are *preserved*, not discretized, and a weighted fuzzy logical combination of those graded biases determines categorization.

A third hint that categorical speech perception is not as categorical as was once thought comes from work by McMurray, Tanenhaus, Aslin, and Spivey (2003). McMurray et al. recorded participants' eye movements while they performed the standard categorical identification task, with sounds from a 'ba'/'pa' voice onset time continuum, by mouse clicking /ba/ and /pa/ icons on a computer screen. Thus, in addition to the record of which icon participants ultimately clicked, there was also a record of when the eyes moved away from the central fixation dot and toward one or another of the response icons while making the categorization. With stimuli near the categorical boundary, the eye-movement record clearly showed participants vacillating their attention between the /ba/ and /pa/ icons. Moreover, despite the identification outcome being identical in the subset of trials categorized as /pa/, the pattern of eye movements revealed substantially more time spent fixating the /ba/ icon when the speech stimulus was near the category boundary in the VOT continuum than when it was near the /pa/ end. These findings point to a clear effect of perceptual gradations in the speech input. The continuous information in the stimulus does not appear to be immediately and summarily thrown away and replaced with some non-decomposable symbol.

Categorical speech perception was arguably the poster child example of the kind of evidence that would be required to defend the Symbolic Approximation Hypothesis; yet its metatheoretical promise has been washed away by a wave of empirical results. In fact, many linguistic phenomena have lost their luster of uniqueness: syntactic processing no longer appears independent of meaning (Tanenhaus & Trueswell 1995); the information flow between language and vision appears to be quite fluid and continuous (Spivey, Tanenhaus, Eberhard, & Sedivy 2002; Spivey, Tyler, Eberhard, & Tanenhaus 2001; see also Lupyan & Spivey, in press); perceptual and motor areas of the brain are conspicuously active during purely linguistic tasks (Pulvermüller 2002). Even the innate perceptual biases that might underlie language acquisition are being re-framed as a developmental penchant for picking up hierarchical structure in *any* time-dependent signal, such as complex motor movement, music, etc. (Hauser, Chomsky, & Fitch 2002; Marcus, Vouloumanos, & Sag 2003; see also Elman, Bates, Johnson, Karmiloff Smith, Parisi, & Plunkett 1996; Lashley 1951; Tallal, Galaburda, Llinás, & von Euler 1993). So maybe language isn't that 'special' after all.

## 2.4 Language is spatial

Rather than language being an independent specialized module, informationally encapsulated from the rest of perception and cognition (Fodor 1983), perhaps it is a process that emerges from the interaction of multiple neural systems cooperating in real time (e.g. Elman et al. 1996; Pulvermüller 2002). If these neural systems are interfacing with one another so smoothly, might they be using a common informational currency? One reasonably likely candidate might be topographic maps, given their prevalence throughout the brain. And if linguistic mental entities exist in some kind of two-dimensional arena of representation, it is natural to expect them (a) to *be located* in particular positions in that two-dimensional space, and also (b) to *subtend*, or 'take up', some portion of that two-dimensional space.

The remaining sections in this chapter describe a range of experiments that reveal these two spatial properties in language-related tasks. When linguistic input instigates the construction of mental representations in this internal topographic space, it elicits eye movements to corresponding locations of external space. Moreover, the shape of the space subtended by these internal representations (e.g. vertically or horizontally elongated) shows consistent agreement in metalinguistic judgements, as well as systematic interference with competing visual inputs in the same regions of space, and these shapes can aid memory when visual cues are compatibly arranged.

## 2.5 Mental models and language comprehension

A great deal of work has revealed evidence for the construction of rich internal representations (or mental simulations) of scenes, objects, and events as a result of comprehending language (e.g. Johnson-Laird 1983; 1998; Bower & Morrow 1990; Zwaan 1999). When someone tells you about the new house they bought, you might feel like you build some kind of image in your 'mind's eye'. This image can be dynamic, as the view changes along with the different aspects of the house being described. And this dynamic mental model can also be interfaced with real visual input. For example, imagine looking at a still photograph of a child's birthday party while a grandparent tells the story of how the children got cake all over themselves. In this case, your dynamic imagery gets overlaid on top of the real image in the photograph.

Altmann and Kamide (2004) showed concrete evidence for this kind of overlay of a dynamic mental model on a viewed scene by tracking people's eye movements while they listened to spoken stories and viewed corresponding scenes (cf. Cooper 1974; Tanenhaus, Spivey-Knowlton, Eberhard, & Sedivy

1995). In one of their experiments, participants viewed line drawings of two animate objects and two inanimate objects for five seconds, e.g. a man, a woman, a cake, and a newspaper. Then the display went blank, and the participants heard a sentence like 'The woman will read the newspaper'. Upon hearing 'The woman', participants conspicuously fixated (more than any other region) the blank region that used to contain the line drawing of the woman. Then, upon hearing 'read', they began fixating the region that had contained the newspaper, more so than any other region. Thus, the memory of the previously viewed objects maintained its spatial organization, and the internal mental model (with its corresponding spatial arrangement) elicited eye movements to appropriate external blank regions of the display when their associated objects were inferable from the content of the speech stream.

Altmann and Kamide's (2004) next experiment demonstrates how we might overlay a *dynamic* mental model onto the static viewed scene. Participants viewed a scene containing line drawings of a wine bottle and a wine glass below a table, and heard 'The woman will put the glass on the table. Then, she will pick up the wine, and pour it carefully into the glass.' In this situation, the mental model must change the spatial location of some of the objects, but the line drawing that is being viewed does not change. Compared to a control condition, participants conspicuously fixated the table (the *imagined* new location of the glass), while hearing 'it carefully', even though the word 'table' was not in that sentence. This finding is consistent with the idea that an internal spatial mental model, constructed from linguistic input, can be interactively 'overlaid' onto an actual visual scene, and thus internally generated images and afferent visual input are coordinated in a two dimensional spatial format of representation that elicits corresponding eye movements.

## 2.6 Topographic spatial representations and verbal recall

Using eye movements to mark and organize the spatial locations of objects and events is a behavior observed in a number of natural situations (cf. Ballard, Hayhoe, & Pelz 1995; O'Regan 1992; see also Spivey, Richardson, & Fitneva 2004). For example, in a series of eye-tracking experiments examining how people tend to exploit spatial locations as 'slots' for linguistically delivered information, Richardson and Spivey (2000) presented four talking heads in sequence, in the four quadrants of the screen, each reciting an arbitrary fact (e.g. 'Shakespeare's first plays were historical dramas. His last play was *The Tempest*') and then disappearing. With the display completely blank except for the lines delineating the four empty quadrants, a voice from the computer delivered a statement concerning one of the four recited facts, and participants were instructed to verify the statement as true or false (e.g. 'Shakespeare's first play was *The Tempest*').

While formulating their answer, participants were twice as likely to fixate the quadrant that previously contained the talking head that had recited the relevant fact than any other quadrant. Despite the fact that the queried information was delivered auditorily, and therefore could not possibly be visually accessed via a fixation, something about that location drew eye movements during recall. Richardson and Spivey (2000) suggested that deictic spatial pointers (e.g. Ballard, Hayhoe, Pook, & Rao 1997; Pylyshyn 1989; 2001) had been allocated to the four quadrants to aid in sorting and separating the events that took place in them. Thus, when the label of one of those pointers was called upon (e.g. 'Shakespeare'), attempts to access the relevant information were made both from the pointer's address in the external environment and from internal working memory.

Richardson and Spivey (2000: experiment 2) replicated these results using four identical spinning crosses in the quadrants during delivery of the facts, instead of the talking heads. Participants seemed perfectly happy to allocate pointers to the four facts in those four locations, even when spatial location was the only visual property that distinguished the pointers. Moreover, in the 'tracking' condition (Richardson & Spivey 2000: experiment 5), participants viewed the grid through a virtual window in the center of the screen. Behind this mask, the grid itself moved, bringing a quadrant to the center of the screen for fact presentation. Then, during the question phase, the mask was removed. Even in this case, when the spinning crosses had all been viewed in the center of the computer screen, and the relative locations of the quadrants implied by translation, participants continued to treat the quadrant associated with the queried fact as conspicuously worthy of overt attention. In fact, even if the crosses appear in empty squares which move around the screen following fact delivery, participants spontaneously fixate the square associated with the fact being verified (Richardson & Kirkham 2004: experiment 1). Thus, once applied, a deictic pointer—even one that attempts to index auditorily delivered semantic information—can dynamically follow its object to new spatial locations in the two dimensional array (e.g. Kahneman, Treisman, & Gibbs 1992; Scholl & Pylyshyn 1999; see also Tipper & Behrmann 1996).

## 2.7 Topographic spatial representations and imagery

Evidence for topographically arranged mental representations is especially abundant in the visual imagery literature (see Denis & Kosslyn 1999 for a review). Reaction times in mental scanning experiments have shown that people take longer to report on properties of imagined objects that are far away from their initial focus point than when the queried object is near the initial focus point (Kosslyn, Ball, & Reiser 1978). This behavioral evidence for mental images exhibiting the same metric properties as real two-dimensional spaces is bolstered by neuroimaging evidence for mental imagery activating some of the topographic neural maps of visual cortex (Kosslyn et al. 1995). Thus, it appears that the same topographical representational medium that is used for afferent visual perception is also used for internally generated visual images.

In fact, as hinted by Altmann & Kamide's (2004) experiments, such internally generated mental images can even trigger eye movements of the kind that would be triggered by corresponding real visual inputs. That is, people use their eyes to *look* at what they are *imagining*. Spivey and Geng (2001; see also Spivey, Tyler, Richardson, & Young 2000) recorded participants' eye movements while they listened to spoken descriptions of spatiotemporally dynamic scenes and faced a large white projection screen that took up most of their visual field. For example, 'Imagine that you are standing across the street from a 40 story apartment building. At the bottom there is a doorman in blue. *On the 10th floor, a woman is hanging her laundry out the window. On the 29th floor, two kids are sitting on the fire escape smoking cigarettes. On the very top floor, two people are screaming.*' While listening to the italicized portion of this passage, participants made reliably more upward saccades than in any other direction. Corresponding biases in spontaneous saccade directions were also observed for a downward story, as well as for leftward and rightward stories. (A control story, describing a view through a telescope that zooms in closer and closer to a static scene, elicited about equal proportions of saccades in all directions.) Thus, while looking at ostensibly nothing, listeners' eyes were doing something similar to what they would have done if the scene being described were actually right there in front of them. Instead of relying solely on an internal 'visuospatial sketchpad' (Baddeley 1986) on which to illustrate their mental model of the scene being described, participants also recruited the external environment as an additional canvas on which to depict the spatial layout of the imagined scene.

Results like Spivey and Geng's (2001; see also Antrobus, Antrobus, & Singer 1964; Brandt & Stark 1997; Demarais & Cohen 1998; Laeng & Teodorescu 2002) provide a powerful demonstration of how language about things that are not visually present is interfaced with perceptual motor systems that treat the linguistic referents *as if they were present*. As a result, a person's eye movements can virtually 'paint' the imagined scene onto their field of view, fixating empty locations in space that stand as markers for the imagined objects there.

Importantly, just as a set of internal representations that are generated by linguistic input can be *located* in particular positions in a two-dimensional space, the next sections examine how these internal representations may also *subtend* some portion of that two-dimensional space; that is, they exhibit a shape that *takes up* some of that space.

## 2.8 Image schemas in metalinguistic judgements

It may seem reasonable to claim that concrete objects and events are cognitively internalized by representations that preserve the metric spatial properties of those objects and events, but how does a topographical, perceptually inspired account of cognition deal with abstract thought? If we do not have amodal symbolic representations for abstract concepts, such as 'respect' and 'success', and instead have topographic representations that are somehow constructed out of some of the early sensory impressions that are associated with those concepts while they were being learned (cf. Mandler 1992), then it is highly unlikely that we would all have exactly the same representations for each abstract concept. While every child's perceptual experience of concrete objects such as chairs, books, cats, and dogs is relatively similar, this is much less so for the perceptual experiences connected with the 'experience' of abstract concepts. Nonetheless, certain basic properties of those representations may be shared by most people within a given culture. For example, if the representations have a shape that is spatially laid out in two dimensions, then perhaps the vertical or horizontal extent of those representations will exhibit a conspicuous commonality across different people. Perhaps the concept of respect, or veneration, is typically first applied to individuals who are *taller* than oneself—hinting at a vertical shape to the topographical representation of a respecting event. Similarly, the concept of success, or winning, is perhaps first learned in competitive circumstances that require getting *above* one's opponents, i.e. another vertical spatial arrangement.

A number of linguists and psychologists have claimed that there is a spatial component to language. The motivations for this claim include capturing subtle asymmetries and nuances of linguistic representation in a schematic spatial format (Langacker 1987; 1990; Talmy 1983), explaining the infant's development from sensorimotor to cognitive reasoning (Mandler 1992), the difficulties in implementing a purely amodal, symbolic system (Barsalou 1999), and a more general account of the mind as an embodied, experiential system (Gibbs 2006; Lakoff & Johnson 1999). Although they are construed differently by various theorists, there appears to be a good case for the conclusion that, at some level, image schemas represent 'fundamental, persuasive organizing structures of cognition' (Clausner & Croft 1999). If so, one would expect a consistent pattern of image schemas to be produced not just by trained linguists and psychologists, but also by naïve subjects.

Controlled psychological experiments have documented the mapping between subjects' spatial linguistic terms and their mental representation of space (e.g. Carlson Radvansky, Covey, & Lattanzi 1999; Hayward & Tarr 1995; Newcombe & Huttenlocher 2000; Regier 1996; Schober 1995). Although there are consistencies in the ways in which spatial language is produced and comprehended (Hayward & Tarr 1995), the exact mapping appears to be modulated by such factors as visual context (Spivey et al. 2002), the common ground between conversants (Schober 1995), and the functional attributes of the objects being described (Carlson Radvansky et al. 1999).

It is probably natural to expect linguistic representations to have at least some degree of topography in their representational format when their content refers directly to explicit spatial properties, locations, and relationships in the world. Spatial language terms appear to be grounded, at least to some extent, in perceptual (rather than purely amodal) formats of representation. In modeling the acceptability judgements for examples of the spatial term 'above', Regier and Carlson (2001) found that the best fit to the data was provided by a model that was independently motivated by perceptual mechanisms such as attention (Logan 1994) and population coding (Georgopoulos, Schwartz, & Kettner 1986). However, an important component of the work presented herein involves testing for this representational format in an arena of language that does *not* exhibit any literal spatial properties: abstract verbs (such as 'respect' and 'succeed'). Work in cognitive linguistics has in fact argued that many linguistic and conceptual representations (even abstract ones) are based on metaphoric extensions to spatially laid out image schemas instead of discrete logical symbols (Gibbs 1996; Lakoff 1987; Langacker 1987; Talmy 1983). This work suggests that if consistency across subjects is observed for spatial depictions of *concrete* verbs, then one should also expect a similar consistency for *abstract* verbs.

A range of studies have pointed out a consistency among speakers in the visual imagery elicited by certain ideas and concepts. For example, Lakoff (1987) offers anecdotal evidence that when asked to describe their image of an idiom such as 'keeping at arms length', people have a considerable degree of commonality in their responses, including details such as the angle of the protagonist's hand. Similarly, Gibbs, Ström, and Spivey-Knowlton (1997) carried out empirical work querying subjects about their mental images of proverbs such as 'A rolling stone gathers no moss' and found a surprising degree of agreement—even about fine details such as the stone bouncing slightly as it rolled.

The approach we take here extends beyond the explicit visual properties of a concept, toward a more schematic spatial representation of verbs. Barsalou's (1999) perceptual symbol system theory endorses the view held by several theorists (Gibbs 1996; Lakoff 1987) that to some degree abstract concepts are represented by a metaphoric extension to more concrete domains. For example, Lakoff has argued that the concept of 'anger' draws on a concrete representation of 'liquid in a container under pressure'.

There is ample evidence to suggest that spatial information plays an important role in many aspects of language processing, from prepositional phrases (Regier & Carlson 2001) to conceptual metaphors (Lakoff 1987). However, the cognitive domains of language and space may have a particularly special 'point of contact' at the level of lexical representation. If we accept the idea that there is a spatial or perceptual basis to the core representation of linguistic items, it would be reasonable to assume that there is some commonality between these representations across different speakers of the same language, since most of us experience rather similar environments, have similar perceptual systems, and by and large communicate successfully. Therefore, we might expect that there would be a consensus among subjects when we ask them to select or draw schematic diagrams representing words. Theorists such as Langacker (1987) have produced large bodies of diagrammatic linguistic representations, arguing that they are constrained by linguistic observations and intuitions in the same way that 'well-formedness' judgements inform more traditional linguistic theories. One approach would be to add to this body of knowledge by performing an analysis of a set of words using the theoretical tools of cognitive linguistics. However, it remains to be seen whether naïve subjects share these intuitions and spatial forms of representation. Therefore, in the same way that psycholinguists use norming studies to support claims of preference for certain grammatical structures, Richardson, Spivey, Edelman, and Naples (2001) surveyed a large number of participants with no linguistic training to see if there was a consensus among their spatial representations of words.

Richardson et al. (2001) empirically tested the claim that between subjects there is a coherence to the imagistic aspects of their linguistic representations. To this end, they addressed two questions: (1) Do subjects agree with each other about the spatial components of different verbs? and (2) Across a forced-choice and an open-ended response task, are the same spatial representations being accessed? It would be of further interest if the subjects' diagrams resembled those proposed by theorists such as Langacker (1987). However, as with more standard norming studies, the real value of the data was in generating prototypical representations that could be used as stimuli for subsequent studies of *online* natural language comprehension.

Richardson et al. (2001) first collected force-choice judgments of verb image schemas from 173 Cornell undergraduates. Thirty verbs were divided into high and low concreteness categories (based on the MRC psycholinguistic database: Coltheart 1981), and further into three image schema orientation categories (vertical, horizontal, and neutral). This latter division was based solely on linguistic intuitions, and as such proved somewhat imperfect, as will be shown later. The verbs were inserted into rebus sentences with circles and squares as the subjects and objects. The participants were presented with a single page, containing a list of the 30 rebus sentences (e.g. [circle] pushed [square], [circle] respected [square], etc.) and four pictures, labelled A to D. Each picture contained a circle and a square aligned along a vertical or horizontal axis, connected by an arrow pointing up, down, left, or right. For each sentence, subjects were asked to select one of the four sparse images that best depicted the event described by the rebus sentence.

The results showed consistent image schematic intuitions among the naïve judges. All 10 of the horizontal verbs had a horizontal image schema as their majority selection, and all but one of the vertical verbs (*obey* was the exception) had a vertical image schema as their majority selection. As it turned out, the neutral group was actually more of a mixed bag of horizontals and verticals, rather than a homogeneously non-spatially biased set of verbs (so much for the experimenters' trained linguistic intuitions). As one quantitative demonstration of consistency among subjects, the particular image schema that was most popular, for any given verb on average, was chosen by 63% of the subjects (with the null hypothesis, of course, being 25% for each option). The second most popular was chosen by 21%, the third by 10%, and the fourth by 5%.

Richardson et al. (2001) also calculated an index for the mean orientation of the primary axis of each verb's image schema, collapsing the leftward and rightward images into one 'horizontal' category, and the upward and downward images into one 'vertical' category. The leftward and rightward image schemas were assigned an angle value of 0, and the upward and downward image schemas an angle value of 90. An average 'axis angle' between 0 and 90 was calculated, weighted by the proportion of participants who selected the vertical and horizontal orientations of image schemas. The five concrete vertical verbs produced an overall mean axis angle of 81°, while the five concrete horizontal verbs produced an overall mean axis angle of 10°. Similarly, albeit less dramatically, the five abstract vertical verbs produced an overall mean axis angle of 55°, while the five abstract horizontal verbs produced an overall mean axis angle of 25°. Item-by-item axis angles can be found in Spivey, Richardson, and Gonzalez-Marquez (2005).

The results of this forced-choice experiment are encouraging for proponents of an image-schematic infrastructure supporting language. However, it could be argued that the pattern of results in this experiment mainly reflects the artificial and limited nature of the forced choice task, in which the restricted and conspicuous set of given image schema choices could be accused of 'leading the witness', as it were.

In their next experiment, Richardson et al. (2001) removed the constraints of a forced choice among a limited set of options, and allowed subjects to create their own image schemas in an open-ended response task. Participants were asked to create their own representation of the sentences using a simplified computer-based drawing environment. The aim was to elicit sparse schematic drawings of the events referred to in the rebus sentences. A custom computer interface allowed Richardson et al. to limit the participants to using a few different circles, a few different squares, and a few extendable and freely rotated arrows. On each trial, the bottom of the screen presented a rebus sentence (using the same verbs from the forced choice experiment), and the participant spent about a minute depicting a two-dimensional rendition of it with the few simple shapes at their disposal. Participants were simply instructed to 'draw a diagram that represents the meaning of the sentence'. When they finished a diagram, they clicked a 'done' button and were presented with the next rebus sentence and a blank canvas.

With a few exceptions, most of the 22 participants attempted to represent the verbs schematically, rather than pictorially (e.g. making humanoid figures out of the circles and arrows). Similar to the 'axis angle' computed in the previous experiment, Richardson et al. (2001) used the coordinates of objects within the canvas frame to define the 'aspect angle' as a value between 0 and 90 to reflect the horizontal versus vertical extent of each drawing. If the objects were perfectly aligned on a horizontal axis, the aspect angle would be 0, whereas if the objects were perfectly aligned on a vertical axis, the aspect angle would be 90. Details, and item-by-item data, can be found in Spivey et al. (2005). Richardson et al. used this measure because they were primarily interested in the horizontal versus vertical aspect of each drawing, and less so in its directionality.

The five concrete vertical verbs produced an overall mean aspect angle of 55°, while the five concrete horizontal verbs produced an overall mean aspect angle of 29°. Similarly, but squashed toward horizontality, the five abstract vertical verbs produced an overall mean aspect angle of 36°, while the five abstract horizontal verbs produced an overall mean aspect angle of 13°. As with the results with the forced-choice response, the 'neutral' verbs behaved more like a mixture of vertical and horizontal verbs.

Despite the free-form nature of the task, there was a reasonably high degree of agreement between participants. Moreover, there was also considerable cross-experiment reliability between the forced-choice experiment and the drawing experiment. By comparing each verb's mean axis angle in the first experiment to its mean aspect angle in the second experiment, via a pointwise correlation, Richardson et al. (2001) found that there was considerable item-by-item consistency between the forced-choice results and the free-form drawing results, with a robust correlation between mean axis angle and mean aspect angle for the verbs in the two tasks; r = 0.71, p < .0001. Importantly, the correlation was statistically significant for all the abstract verbs alone (r = .64, p < .0001), as well as for all the concrete verbs alone (r = .76, p < .0001). Thus, the two measures (forced-choice and free-form drawing) appear to be accessing the same internal representations, i.e. image schemas that are relatively stable across the different tasks and across different subjects.

These findings provide compelling support for the image schematic approach to language endorsed by cognitive linguistics (e.g. Langacker 1987; Talmy 1983; see also Coulson & Matlock 2001). However, there exist some informative cases where the experimenters' 'trained linguistic intuitions' were refuted by the participants. For example, in the neutral condition, both 'perched' and 'rested' were consistently given a vertical interpretation by participants in both tasks. Additionally, the average image schema for 'obeyed' was considerably more horizontal than had been expected. These observations highlight the importance of using normative methodologies from psychology to accompany traditional introspective methodologies from linguistics (cf. Gibbs & Colston 1995).

The results described here could be taken as further evidence challenging the 'classical' view that linguistic representations are amodal, symbolic entities (e.g. Marcus 2001; Dietrich & Markman 2003). Alternatively, one could maintain that all we have shown is that such hypothetical, amodal representations have easy access to spatial information in a way that is consistent across users of a language. Given that language is learned and used in a spatially extended world that is common to all of us, then of course participants will find consistent relations between certain spatial dimensions and certain words. This could happen whether the underlying linguistic representations were multi-modal 'perceptual simulations' (Barsalou 1999) or amodal entries in a symbolic lexicon. Thus, the spatial consistency revealed by metalinguistic judgements may not be inherent to *linguistic* representations, but instead may be part of some other body of knowledge that can be deliberatively accessed via an amodal lexical entry.

What is required is a measure of language processing that does not involve metacognitive deliberation. If these kinds of spatial representations become active during normal real-time comprehension of language, and can be revealed in a concurrent but unrelated perceptual task, then it becomes much more difficult to argue that they are secondary representational appendices, separate from the core linguistic symbols, that are merely strategically accessed when some psychology experimenter overtly requests them.

## 2.9 Image schemas in perception

The spatial representations that Richardson et al.'s (2001) participants ascribed to verbs could be part of the metaphoric understanding that underlies much of our language use, and may be rooted in embodied experiences and cultural influences (e.g. Gibbs 1996; Lakoff 1987). Alternatively, perhaps these spatial elements are more like idioms, or linguistic freezes—historical associations that are buried in a word's etymology but are not part of our core understanding of the concept (Murphy 1996). This issue forms the central question of the next set of experiments to be described. Are the spatial representations associated with certain verbs merely vestigial and only accessible metacognitively, or are they automatically activated by the process of comprehending those verbs?

Richardson, Spivey, Barsalou, & McRae (2003) operationalized this question by presenting participants with sentences and testing for spatial effects on concurrent perceptual tasks. An interaction between linguistic and perceptual processing would support the idea that spatially arrayed representations are inherent to the conceptual representations derived from language comprehension (e.g. Barsalou 1999; see also Kan, Barsalou, Solomon, Minor, & Thompson Schill 2003; Solomon & Barsalou 2001; Zwaan, Stanfield, & Yaxley 2002). The interactions predicted were specific to the orientation of the image schema associated with various concrete and abstract verbs. Richardson and colleagues used the empirically categorized set of verbs from the norming studies of Richardson et al. (2001). Because it was assumed that image-schematic spatial representations bear some similarity to visuospatial imagery (albeit a weak or partially active form), they predicted that it would interact with perceptual tasks in a similar fashion.

Evidence of visual imagery interfering with visual perception was discovered at the turn of the century (Kuelpe 1902; Scripture 1896), and rediscovered in the late 1960s (Segal & Gordon 1969). In demonstrations of the 'Perky effect' (Perky 1910), performance in visual detection or discrimination is impaired by engaging in visual imagery. Imagery can also facilitate perception when there is a relatively precise overlap in identity, shape, or location between the imaginary and the real entity (Farah 1989; Finke 1985). In the more general case of generating a visual image and detecting or discriminating unrelated stimuli, imagery impairs performance (Craver Lemley & Arterberry 2001). Richardson et al.'s (2003) first experiment tested the hypothesis that axis-specific imagery activated by verb comprehension would *interfere* with performance on a visual discrimination task.

In this dual-task experiment, 83 participants heard and remembered short sentences, and identified briefly flashed visual stimuli as a circle or square in the upper, lower, left, or right sides of the computer screen. The critical sentences contained the 30 verbs for which Richardson et al. (2001) had collected image schema norms. The data from those two norming tasks were combined and the result used to categorize the verbs empirically as either horizontal or vertical (instead of relying on experimenters' intuitions). Richardson et al. (2003) predicted an interaction between the linguistic and visual tasks. That is, after comprehending a sentence with a vertical verb, and presumably activating a vertically extended image schema in some spatial arena of internal representation, participants' perception would thus be subtly inhibited when an unrelated visual stimulus appeared in the top or bottom locations of the screen. Likewise, after a horizontal verb, perception in the left and right positions should be inhibited.

Reaction time results showed a reliable interaction between verb orientation and position of the visual stimulus. When the verb's image schema was vertically elongated (e.g. 'respect', 'fly'), reaction times were 15 ms *slower* to stimuli presented along the vertical meridian (534 ms) than to stimuli presented along the horizontal meridian (519 ms). Conversely, when the verb's image schema was horizontally elongated (e.g. 'give', 'push'), reaction times were 7 ms *faster* to stimuli presented along the vertical meridian (516 ms) than to stimuli presented along the horizontal meridian (523 ms). Interactions with concreteness did not approach significance in this study, suggesting that this result was not significantly different for concrete and abstract verbs. Admittedly, these effects are subtle, but the interaction is statistically robust, and the effects have been replicated in another laboratory (Bergen, Narayan, & Feldman 2003); however, the abstract verbs may be less effective than the concrete verbs at generating this spatial imagery (Bergen, Lindsay, Matlock, & Narayanan 2007).

Since these verbs modulated online perceptual performance in a spatially specific manner predicted by the norming data, this suggests that Richardson et al.'s (2001) results were not an artefact of offline tasks that require deliberate spatial judgements. More importantly, this result provides evidence that

comprehending a spoken verb automatically activates a visuospatial representation that (in its orientation of the primary axis, at least) resembles the image
schema associated with the meaning of that verb. Crucially, one should not
conceive of the spatial representation activated by these verbs as a kind of raw
spatial priming (or a vertically/horizontally shaped attentional window). The
effect being found is of *interference* between the activated image schema and
the afferent visual input, so the spatial representation activated by the verb in
this topographic arena clearly has an identity of its own that causes it to be
incompatible with the unrelated visual stimulus coming in.

## 2.10  Image schemas in memory

In a second experiment with the same set of sentences, Richardson et al. (2003)
investigated how language comprehension interacts with a memory task. It has
been robustly shown that imagery improves memory (Paivio, Yuille, & Smythe
1966). Also, visual stimuli are remembered better when they are presented in
the same spatial locations at presentation and test (Santa 1977). Thus, it was
hypothesized that spatial structure associated with a verb would influence the
encoding of concurrent visual stimuli, which could then be measured later
during retrieval.

During each block of study trials, 82 participants heard six sentences while
line drawings of the corresponding agent and patient were presented sequen-
tially in the center of the screen. During the test phase, the pictures were presented simultaneously in either a horizontal arrangement (side by side) or
vertical arrangement (one above the other). Participants were instructed to
indicate by button-press whether the two pictures had been shown together as
part of a sentence or not. In half of the test trials, the two pictures were taken
from different sentences; in the other half (the critical trials), the pictures were
from the same study sentence. It was predicted that the picture pairs would
later be recognized faster if they were presented in an orientation consistent
with the verb's image schema.

As predicted, memory was facilitated when the test stimulus orientation
and the verb orientation coincided. When the recall test images were arranged
along the vertical meridian, reaction times for recall of a vertical verb (1,299 ms),
were 97 ms *faster* than for recall of a horizontal verb (1,396 ms). Conversely,
when the recall test images were arranged along the horizontal meridian, reaction times for recall of a vertical verb (1,289 ms), were 16 ms *slower* than for
recall of a horizontal verb (1,273 ms). Interactions with concreteness did not
approach significance, suggesting that the effect is about the same for concrete
and abstract verbs.

Thus, verb comprehension influenced how visual stimuli were encoded in memory, in that recognition times were faster when the stimuli were tested in an orientation congruent with the verb's image schema. In contrast to the interference effect found in visual perception, image schemas facilitated performance in this memory task. One interpretation is that during study, verb comprehension activated an image schema, and the spatial element of this image schema was imparted to the pictures, as if the verb's image schema were acting as a scaffold for the visual memory. The pictures were then encoded in that orientation, and hence identified faster when presented at test in a congruent layout (e.g. Santa 1977).

This pair of findings on verbal image schemas affecting perception and memory constitutes persuasive evidence for topographically arranged representational contents being automatically activated as core components of linguistic meaning. In addition to language influencing visual perception, the reverse, where visuomotor processing influences language, appears to work as well. Toskos, Hanania, and Hockema (2004) showed that vertical and horizontal eye movements can influence memory for these vertical and horizontal image-schematic verbs. Finally, syntax plays a role here as well. Results from a pilot study, with an offline forced choice similar to Richardson et al.'s (2001) experiment 1, indicate that placing a verb in different syntactic frames can alter the orientation of the image schema's primary axis. For example, although naive participants tend to select a vertically arranged image schema for a sentence like 'The circle respected the square', they tend to select a horizontally arranged image schema for a sentence like 'The circle and the square respected each other'.

## 2.11  General discussion

In this chapter, we have shown how mental models (Altmann & Kamide 2004), verbal memory (Richardson & Spivey 2000), visual imagery (Spivey & Geng 2001), and even the online comprehension of spoken verbs (Richardson et al. 2003; see also Matlock 2004) involve the activation of representations that are *located in specific positions* in a topographical space and that *subtend specifically shaped regions* of that space. We have argued that a great deal of internal cognitive processing is composed of these continuous partially overlapping mental entities that exist in a two-dimensional topographic medium of representation, rather than binary symbols operated on by logical rules.

More and more, it is beginning to look as though discrete and static non-overlapping symbolic representations are not what biological neural systems use (cf. Georgopoulos 1995). Recently, the field has been witnessing a substantial accrual of behavioral evidence for cognition using analog spatial formats of representation that are nicely compatible with the topographical formats of representation known to exist in many areas of the human brain (e.g. Churchland & Sejnowski 1992; Swindale 2001). Therefore, to the degree that one believes that the workings of the brain are of central importance to the workings of the human mind, it should be clear that the mind cannot be a digital computation device.

In philosophy of mind, functionalism argues that, since 'mind' is best defined as *the causal relationships* between states of the system, the actual physical material on which those states and relationships are implemented is irrelevant. While discouraging inquiry into neuroscience, or by taking neuroscience to be of secondary importance to a functional characterization of cognition, these functionalists duck the label of 'Cartesian dualist' by acknowledging that the physical matter of the brain is indeed the key subsystem underlying human mental activity (perhaps along with the body and its environment)—it's just that this fact is irrelevant to understanding how the *mind* works. However, there is at least one self-contradiction hiding in this juxtaposition of beliefs (see also Kim 1998). If the brain can be understood as a complex dynamical system, which is almost certainly true, then it is likely to exhibit sensitivity to initial conditions. That is, extremely subtle aspects of its state—the equivalent of significant values in the tenth decimal place, if you will—can have powerful long-lasting effects on where in state space the system winds up many time steps later. As observed in symbolic dynamics, when these subtle 'tenth decimal place' properties are ignored (or rounded off) in a complex dynamical system, disastrous irregularities and violations can result (cf. Bollt et al. 2000). Unfortunately, ignoring them is exactly what the Symbolic Approximation Hypothesis encourages us to do.

Nonetheless, digital symbol manipulation as a metaphor for how the mind works was a productive simile for the first few decades of cognitive science. It just may be outliving its usefulness, and beginning to inhibit, rather than facilitate, genuine progress. According to the Symbolic Approximation Hypothesis, even if high-level neural systems in the brain cannot actually construct true Boolean symbols and implement genuine discrete state transitions, what the neural patterns are doing in those areas may be *close enough* to actual symbolic computation that the inaccuracies are insignificant when we approximate those biological processes with artificial rules and symbols. This is the risky wager against which Rosen (2000) warns in the quotation at the beginning of this chapter. The many different results described in this chapter point to a broad range of evidence supporting the existence of topographically arranged spatial representations, in language, imagery, and memory, that are substantially incompatible with a symbolic, propositional account of mental activity. As more and more of cognition turns out to be using these continuous representational formats, instead of discrete logical formats, the symbolic approach to cognition may eventually find itself out of a job.

Of course, reporting evidence for internal mental constructs that appear to be composed of a spatial format of representation does not, by itself, prove that symbolic propositional representations do not exist. The two types of representation could, in principle, coexist (Sloman 1996). However, as more and more of the 'textbook examples' of discrete amodal symbolic thought give way to more successful continuous, perceptually grounded, and dynamic accounts, e.g. visual imagery (Kosslyn et al. 1995), conceptual knowledge (Barsalou 1999), categorical perception (McMurray et al. 2003), and even language (Richardson et al. 2001 2003), one has to wonder when this succession of lines in the sand will cease.

At this point, the progressively advancing movement of dynamical systems approaches to cognition appears to be the most promising candidate for a framework of cognition (cf. Elman et al. 1996; Kelso 1994; Port & van Gelder 1995; Spivey 2007; Thelen & Smith 1994; Van Orden, Holden, & Turvey 2003; Ward 2002). The dynamical systems framework naturally accommodates both spatial and temporal continuity in the representational format of perception and cognition (Spencer & Schöner 2003). As this kind of continuity in representational formats becomes broadly recognized in more cognitive phenomena, we predict that the symbol-minded information-processing approach to psychology will give way to a dynamical systems framework.

## Acknowledgements

Much of the work described herein was supported by NIMH grant #R01-63691 to the first author and by a Cornell Sage Fellowship to the second author. The authors are grateful to Linda Smith and the entire workshop membership, as well as Aare Laakso, Mark Andrews, Rick Dale, Eric Dietrich, Monica Gonzalez-Marquez, Ken Kurtz, Ulric Neisser, and Art Markman for helpful discussions and comments that contributed to various ideas presented herein.

# 3

# Spatial Tools for Mathematical Thought

KELLY S. MIX

Of all the aspects of language children have to learn, the words, symbols, and algorithms used to represent mathematical concepts may be the most opaque. To help children grasp mathematical language and the ideas it represents, educators have developed a variety of concrete models, or 'manipulatives'. These objects construe mathematical relations as spatial relations, thereby allowing children to experience these abstract notions directly before learning to describe them symbolically. The use of such materials is widespread. In fact, concrete models form the backbone of many early childhood mathematics curricula (Montessori, Math Their Way, etc.).

Although concrete models have intuitive appeal, there are many questions regarding their effectiveness. Research on the most fundamental question—whether these instructional tools are helpful—has yielded mixed results (Ball 1992; Fennema 1972; Friedman 1978; Fuson & Briars 1990; Goldstone & Sakamoto 2003; Goldstone & Son 2005; Kaminski, Sloutsky, & Heckler 2005; 2006a; 2006b; Moyer 2001; Peterson, Mercer, & O'Shea 1988; Resnick & Omanson 1987; Sowell 1989; Suydam & Higgins 1977; Uttal, Scudder, & Deloache 1997; Wearne & Hiebert 1988). Furthermore, almost no research has addressed *how* or *why* these materials might help (Ginsburg & Golbeck 2004). These two facts may not be coincidental. I will argue that one reason manipulatives do not always look effective is that they play different roles in different situations. To assess the usefulness of these materials, it may be necessary to at least speculate about how and why they help—to identify the underlying mechanisms these materials might engage and evaluate at a more precise level whether they succeed. My aim in this chapter is to outline the possibilities based on recent advances in cognitive science, cognitive development, and learning sciences. In doing so, I hope to resolve some discrepancies in the extant literature, as well as develop a framework for future research on these potentially important spatial tools.

## 3.1 Acquisition of mathematical language: obstacles and opportunities

Language is arguably the centerpiece of human cognition. Words not only allow us to communicate with others, but also focus our attention, organize our memories, and highlight commonalties and relationships we might otherwise overlook (Gentner & Rattermann 1991; Imai, Gentner, & Uchida 1994; Markman 1989; Rattermann & Gentner 1998; Sandhofer & Smith 1999; Smith 1993; Waxman & Hall 1993; Waxman & Markow 1995). Words and symbols also support new insights simply by freeing up cognitive resources (e.g. Clark 1997). In fact, language has been called the ultimate cognitive tool because it scaffolds complex thought so effectively (Clark 1997; Vygotsky 1978).

If language performs these functions for relatively straightforward concepts, such as 'dog' or 'cup', consider the role it must play in mathematical thought. You can directly experience a dog, even if you don't know what to call it. But how do you directly experience something like subtraction with borrowing? Or long division? Naturally occurring examples of these notions are so infrequent, it is unlikely that children could discover them on their own. Indeed, although the developmental origins of number concepts remain in dispute (see Mix, Huttenlocher, & Levine 2002a for a review), there is general agreement that mathematical language is needed to attain all but the most primitive, quasi-mathematical abilities (Carey 2001; Gelman 1991; Mix, Huttenlocher, & Levine 2002b; Spelke 2003; Spelke & Tsivkin 2001; Wynn 1998).

This makes sense given that mathematics is, by definition, a set of formalisms expressed and constructed via symbol manipulation. When we teach children mathematics, we are passing down the body of insights accumulated by scores of mathematicians working with symbols throughout human civilization. The history of mathematics is filled with examples of conceptual advances built on the achievements of preceding generations (Ifrah 1981; Menninger 1969). Our young learners recapitulate these cultural developments in their own development—standing on the shoulders of giants, as it were. And just as mathematicians needed the language of mathematics to discover these insights, children seem to need it to recognize those insights.

However, there is an irony in this codependence between mathematical thought and mathematical language. The same conditions that make symbols and words especially vital for constructing mathematical concepts, whether by cultures or individuals, also renders them especially difficult to acquire. For example, numerous calculation procedures, such as long division, require place value. These procedures formalize real-life situations in which large quantities are combined or partitioned, but do so in a way that is rapid and efficient.

Stop and consider how you might divide 5,202 sheep among 9 farmers without long division. It would be possible, but it would take a while and there would be plenty of room for error. The benefits of place value procedures are clear. The problem, in terms of learning such procedures, is that they are one step removed from physical experience—they are based on direct experience with symbols, not objects. Hence the dilemma: If a concept is inaccessible without certain symbols, how do you learn what the symbols mean?

This problem is similar to the well-known indeterminacy problem in language learning (i.e. Quine 1960), but it is not exactly the same thing. Although it is unclear what aspect of rabbithood you mean when you point to a rabbit and say, 'Gavagai', at least there is a rabbit to point to! It is less obvious how children could get to the idea of 'two tens plus six' after hearing a pile of blocks named 'twenty-six'. Thus, although mathematical language suffers from the same indeterminacy as other language, it has added challenges because mathematical 'objects' are often mental constructions. In short, you can learn the word 'rabbit' by living in the physical world, but you cannot learn what tens and hundreds are without inhabiting a symbol system as well.

This added layer of complexity likely underlies a widely recognized gap between children's intuitive understanding of mathematics and their proficiency with related symbolic procedures (Greeno 1989; Schoenfeld 1987). In many cases, these two kinds of understanding develop independently. For example, children learn to recite the counting sequence separately from learning to match or identify small quantities (Fuson 1988; Mix, Huttenlocher, & Levine 1996; Mix 1999; Shaeffer, Eggleston, & Scott 1974; Wagner & Walters 1982; Wynn 1990). In fact, they can say the count word list with perfect accuracy years before they realize what counting is about (i.e. the last word of the count stands for the cardinality of the set) (Wynn 1990; 1992).

Children often learn to perform place value procedures the same way—mechanically carrying out the procedures as Searle's (1980) translator learned to compose Chinese sentences by following a set of grammatical rules. However, just as it may be possible to generate Chinese sentences without understanding them, children can to carry out mathematical algorithms without knowing what physical realities the algorithms represent. This becomes evident when they fail to notice mistakes that should be obvious based on estimation alone (e.g. '10 + 6 = 2') or have trouble transferring their 'understanding' to new problems (e.g. '3 + 4 + 5 = 7 + _?_ ') (Alibali 1999; Ginsburg 1977; Rittle-Johnson & Alibali 1999). This lack of understanding is also apparent when children carry out symbolic procedures correctly but without connection to informal knowledge or concrete representations (Greeno 1989; Mack 1993; 2000; 2001; Resnick & Omanson 1987; Schoenfeld 1987). And there are yet other situations where children never even achieve accurate computation, instead resorting to guessing or applying known procedures in an incorrect way (e.g. '1/2 + 1/4 = 2/6') (Mack 1993).

## 3.2 Using spatial tools to bridge the gap

To help children connect their intuitive understanding of mathematics to the related symbolic procedures, some educational theorists advocated the use of concrete models (Bruner 1961; Dienes 1960; Montessori 1964). These models are structured so as to embody mathematical relationships more transparently than everyday objects can. The idea is that interacting with such objects provides a stepping stone between everyday experience and abstract formalisms. For example, to teach place value concepts, Montessori developed a set of beads that illustrates the relations among ones, tens, hundreds, and thousands (see Figure 3.1). These objects are different from the objects children are likely to encounter in their day-to-day activities, and in that sense they may be less intuitive. Yet they provide a means of physically manipulating the relations among different place values that is lacking in written symbols.

Other examples of math manipulatives include Cuisinaire rods—blocks that illustrate the decomposition of numbers (i.e. '1 + 6 = 2 + 5 = 3 + 4'),

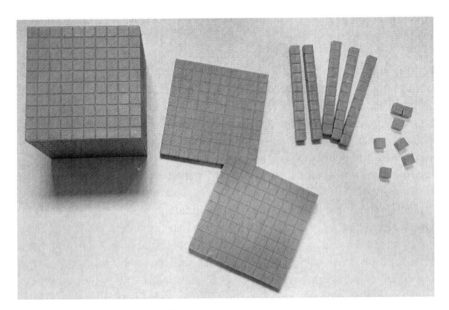

FIGURE 3.1. Materials used to illustrate base-10 relations

and fraction pieces—bars or pie pieces that can be used to compare, add, or subtract fractions. What all these materials have in common is that they are tangible objects explicitly engineered to represent a particular mathematical relation. This is usually accomplished by recasting a mathematical relation in terms of a spatial relation. So, for example, the Cuisinaire rod for 'three' is half the length of the rod for 'six.' This means that when children lay a 'three' and a 'three' together, end-to-end, the result will be exactly as long as a 'six.' In this way, the decomposition of small numbers is explicitly represented using space.

Concrete models are implemented in many different ways based on a range of variation in structure, degree of contact with the materials, symbolic mapping, amount of exposure, and number of instantiations (e.g. Hiebert & Carpenter 1988; Stevenson & Stigler 1992). In terms of structure, experiences can be as unconstrained as free play. One approach is to simply make the materials available so that children can discover the properties of them through unguided exploration. More structured activities might involve playing games or performing computations with the materials. An example would be the Banker's Game, where children trade chips that stand for different place values. To play, a child rolls a die and gets to take the corresponding number of blue chips (i.e. ones) from the banker. Once the child has rolled enough to get at least ten blue chips, he can trade the ten in for one yellow chip (i.e. a ten). The game continues until one of the children wins by accumulating enough yellows (ten) to turn them in for a red chip (i.e. a hundred).

Manipulative activities also vary in terms of children's contact with the materials. At one extreme, children each receive their own set of materials to touch and manipulate. Sometimes children share materials in pairs or small groups, where they alternate between watching their classmates manipulate the objects and manipulating the objects themselves. In some cases, only the teacher has the materials and demonstrates activities as the children watch. For example, place value beads might be used to illustrate subtraction with borrowing (i.e. when you move ten from the tens place to the ones place, you trade in a string of ten beads, or a 'long', for ten individual beads). Concrete models are frequently presented as photographs or schematic drawings in textbooks, workbook pages, or educational software. Here there is no direct contact, but the spatial relations are preserved.

A third dimension that can vary is the relation of manipulatives to written symbols. Many activities, structured or not, can be carried out with no mention of the corresponding written symbols. Children can learn to play the Banker's Game, for example, without ever realizing that the numeral 111 could be represented with three chips—a red, a yellow, and a blue. Alternatively, it would be possible to start with written problems and then use concrete materials to illustrate them. For example, a teacher could show children the algorithm for subtraction with borrowing, let them learn to do it by rote, and then introduce the materials to provide a deeper understanding of this known procedure. Of course, there are many variations in between, including decisions regarding whether to do symbolic and concrete activities in alternation or in tight coupling (i.e. writing down symbols at each stage of a computation using blocks).

Obviously, a fourth dimension of variation involves the amount of exposure children receive. Some concrete models, like fingers, are always available. Classroom manipulatives may be used extensively, occasionally, or rarely. Hypothetically speaking, children could work with these materials for hours every day. In reality, mathematics instruction averages 323 minutes per week, or roughly an hour a day, including time for written practice, teacher instruction, and assessments (McMurrer 2007). Therefore, the total amount of exposure to manipulatives could reduce to minutes per month.

A related dimension is how many manipulatives children are given to illustrate a particular problem or relation. In the United States, it is typical to present multiple instantiations of the same problem. For example, for place value instruction, teachers might use bundled popsicle sticks, beans and bean sticks, place value blocks and/or Montessori beads, colored poker chips, and so forth. This approach is thought to promote abstraction by highlighting ways in which all these differing materials overlap (e.g. Dienes 1960). Of course, this means that children will receive relatively little exposure to each instantiation. In contrast, Asian schools typically use only one model (Stevenson & Stigler 1992). This approach is thought to promote abstraction by deepening children's understanding of the materials or promoting expertise. This also means that children will receive relatively greater exposure to this single instantiation, but may not see how it overlaps with other models or experiences.

Considering all the different ways concrete materials can be implemented, the question of whether these materials work is not as simple as it seems. Each of these variations could work or not work for different reasons based on the cognitive resources required and the underlying mechanisms that might be engaged. In the next section, I outline what these resources and mechanisms might be. Then, I reconsider these implementation issues in light of this new framework.

## 3.3  Why might concrete models help?

The original impetus for using concrete models was the notion of a concrete-to-abstract shift in cognitive development (Bruner 1961; Piaget 1951). The idea was that children, unlike adults, are not capable of purely symbolic thought. Instead, children were supposedly limited to what was directly perceivable—to concrete, in the here-and-now, experience. In fact, children seemed so trapped by their perceptions that they could be led into countless logical traps without even knowing it.

From an instructional standpoint, the solution seemed clear. You can't teach symbolic relations to perception-bound children by having them manipulate symbols. You need, instead, to provide concrete experiences that will impart these understandings at an intuitive level. Once children have achieved the capacity for abstract thought, they would then use this storehouse of targeted experiences as referents for the otherwise opaque mathematical symbols.

But advances in cognitive science and cognitive development have changed the way psychologists think about the concrete-to-abstract shift. In fact, this transition has been attacked from both sides—some arguing that children are capable of abstract thought (e.g. Gelman & Baillargeon 1983) and others arguing that even adult thought relies on sensorimotor experience and perceptual scaffolding (Clark 1997; Port & van Gelder 1995; Thelen & Smith 1994). These advances have also provided a more nuanced description of the processes that underlie learning, symbol grounding, and generalization. From this new perspective, new ideas about the potential roles of concrete models emerge. In the following section, I review four specific mechanisms that concrete materials might engage.

### 3.3.1 *Concrete models might generate actions*

Traditionally, movement was considered a low-level behavior under biological control, which operated separately from abstract, higher level cognition. However, movement is now recognized as a central human behavior and a linchpin for cognitive development and learning. This is because movement and thought are tightly coupled in the continuous flow of experience, and thus should be an inherent part of the memories from which concepts are built (see Clark 1997; Glenberg 1999; Port & van Gelder 1995; Thelen & Smith 1994).

Numerous empirical studies have shown this to be the case. Once infants start to move on their own, they exhibit a range of cognitive advances, such as improved memory for object locations (Acredolo, Adams, & Goodwyn 1984; Bertenthal, Campos, & Barrett 1984; see Thelen & Smith 1994 for a review). Moreover, infants' understanding of visual cliffs and slopes appears to be linked to specific types of movement, such as walking versus crawling (Adolph, Eppler, & Gibson 1993). Actions also influence category development. Smith (2005) demonstrated that 2-year-olds extend novel words differently depending on how they had manipulated the target objects prior to test. For example, if they had moved an object vertically while learning a novel word, they were more likely to extend its name to the same object in an upright position. In adults, there is a clear link between object recognition and movement such that when an object is recognized visually, memory for the actions typically associated with it (such as drinking from a cup) are automatically activated as well (Ellis & Tucker 2000; Creem & Profitt 2001; Glenberg, Robertson, Kaschak, & Malter 2003). Adults remember dialogue and sequences of events when they learn the information with movement (Noice & Noice 2001). Adults learning words in a foreign language remember better if they smile, or make a pulling motion while they learn (Cacioppo et al. 2006).

From this perspective, a clear advantage to teaching mathematics with concrete materials is that these give children something to act upon. If cognition is built from movement through space and memories for these movements, then mathematical cognition must be built from mathematically relevant movements. But how many naturally occurring movements are there for ideas like long division? Children are not likely to stumble upon these concepts in their everyday actions. And there probably is not enough time for them to build such concepts that way, even if they did. Bear in mind that the job of math teachers is to convey the insights achieved by generations of expert symbol manipulators (i.e. mathematicians) to cognitively immature and relatively inexperienced beings who are novice symbol manipulators at best. There is a lot of information for children to digest in a short amount of time. Concrete models may supply a crucial stream of movement information that is targeted to these symbolic procedures, thereby supporting the natural process of learning under the unnatural demands of formal mathematics instruction.

This perspective also yields a new insight into why children might learn written algorithms by rote. Consider, for example, subtraction with borrowing. When children are taught the sequence of written markings for subtraction, their concept of subtraction is tightly coupled with these movements. In fact, it may be hard for them to think about subtraction without making these movements. This might manifest itself when children are asked to solve such problems mentally, without access to paper and pencil. Under such conditions, there may be a strong impulse to gesture as if to write these markings on a table or in the air. And if these actions are children's only bodily experience with subtraction, then they may cling to the written algorithms—whether or not they understand them at a deeper level—because these provide the only opportunity for re-experiencing subtraction through movement. (If this claim seems implausible, try telling someone how to tie a shoe without reconstructing the movements yourself.)

A final implication for mathematics instruction is that the particular movements required by a task or procedure may interfere with learning if they are too complicated. For example, writing a four-digit addition problem with a pencil requires considerable fine motor coordination. Novice writers or children with fine motor delays may be so hung up on making the correct movements that they have difficulty thinking about the problem to be solved. This goes beyond saying that motor demands can be distracting. Instead, the claim is that children who are prevented from moving may also be prevented from learning—that is, they may have difficulty accessing new concepts that are not readily linked to an action.

One approach to dealing with this issue is to provide tiles or slips of paper with numerals written on them. For example, Montessori classrooms provide strips of paper that make explicit the decomposition of large numbers. So, to write the number 348, children would retrieve three strips—one that says '300', one that says '40', and one that says '8'. By stapling these on top of each other, the child effectively 'writes' 348 without needing the ability to print. Furthermore, the movement involved here (i.e. layering strips of paper that each represent a place value) mirrors what is meant by a multi-digit numeral more directly than writing the individual numerals.

### 3.3.2 *Concrete models might generate conceptual metaphors*

Early proponents of concrete models for mathematics assumed that there was a concrete-to-abstract shift in development (e.g. Bruner 1961; Piaget 1951). From this perspective, children move from intuitive reasoning based on direct perception and contextual information to logic reasoning based on fully decontextualized formalisms. However, there is strong evidence that the seemingly abstract reasoning of adults is embodied in concrete perception and action. Yet there are obvious changes in reasoning from childhood to adulthood. Adults are capable of logical thought. They comprehend abstract concepts, like justice, and are less swayed by erroneous perceptual cues. They seem able to generalize across disparate situations in ways that children cannot. How is it possible for an embodied mind to generate such disembodied behaviors?

The answer, according to some cognitive scientists, is that adults do shift toward an emphasis on symbols but that these symbols are grounded in concrete perception and action. Thus, as in previous views of adult cognition, humans can manipulate symbols and complete patterns. However, though these manipulations may be several steps removed from concrete reality, the symbols themselves remain embodied because they originate in connection to concrete experience (Barsalou 1999; Clark 1997; Glenberg 1997; Greeno 1989; Lakoff & Johnson 1980b; Lakoff & Nunez 2000).

Lakoff and Nunez (2000) used this approach to explain the genesis of higher mathematics. They argued that the same processes that allow scholars to invent mathematics also allow children and novices to learn mathematics. Specifically, they proposed that mathematical thought consists of layer upon layer of conceptual metaphors—metaphors that originate from experience in the physical world.

One kind of metaphor, the grounding metaphor, is directly tied to experience with physical objects. For example, most children have extensive experience with containment. They move objects, like blocks and toy cars, in and out of containers. They watch other people move objects in and out of containers (e.g. pouring cereal from the box into a bowl). And they even move themselves into and out of containers (clothing, bedding, tunnels, etc.). This massive experience provides the grounding for the containment metaphor—a notion that underlies a variety of mathematical concepts, such as numerical identity, decomposition, and bounded intervals. Grounding metaphors are thought to be self-evident from experience with objects. Thus, they are internalized spontaneously without formal instruction.

The second kind of metaphor, the linking metaphor, connects one domain of mathematics to another. For example, by connecting geometry to arithmetic, it is possible to conceive of numbers as points on a line. Though these metaphors are not derived directly from concrete experience, they are built from primitives that are (i.e. grounding metaphors and image schemas). It is thought that linking metaphors are not spontaneously discovered, but instead are learned through formal instruction.

In a very general sense, the developmental progression described by Lakoff & Nunez (2000) resembles the concrete-to-abstract shift. Children first have direct interactions with objects. These become internalized as metaphors, and then basic arithmetic and number concepts are acquired with reference to these metaphors. Higher mathematics is built from new metaphors that emerge from mappings among the grounding metaphors. In fact, the idea of metaphors mapping among metaphors is something like Piaget's characterization of abstract thought as 'operating on operations'. The important difference is that, in the view of Lakoff & Nunez, higher mathematics is never completely distinct from concrete experience. Quite to the contrary, the argument is that higher mathematics can be understood only in reference to these experiences, even in adults.

So, how are conceptual metaphors internalized? According to Lakoff & Nunez (2000), these are the product of *conflation*—the simultaneous activation of distinct areas of the brain that are concerned with different aspects of experience. To illustrate, they point out that people construe relationships in terms of warmth, as in 'He gave me the cold shoulder' or 'She has a warm smile'. The connection of temperature to social responsiveness, they argue, arises from the conflation of human contact and body warmth that is literal when one is held as an infant. In other words, a babe in arms experiences her mother's loving gaze while also feeling warm, and conflates these two experiences in memory. Mathematical metaphors are thought to originate via the same mechanism, through the conflation of certain experiences with early enumeration processes. For example, children walking across the room, or up a staircase, would directly experience segments in a path. If they do so while simultaneously enumerating them, the conflation of these experiences could set children up to see numbers as points on a line.

From this perspective, in which mathematical concepts are built from perceptually grounded metaphors, concrete models for mathematics instruction could play several possible roles. The most obvious is that they could provide fodder for the creation of new grounding metaphors. Math manipulatives are designed to explicitly represent mathematical relations. Perhaps by interacting with these materials, children have experiences that they would not typically have in play with everyday objects, thereby leading to conceptual metaphors that would not normally arise. Because these grounding metaphors would be tailored to notions that underlie higher mathematics, they could be particularly valuable later in development.

For example, when children manipulate place value blocks, their memories of physically constructing and decomposing sets could become conflated with their visual memories of the way different groupings compare to one another (e.g. ten unit blocks laid end-to-end look just like a 'long' that cannot be taken apart). Because these materials explicitly represent base-10 structure, they virtually force children to see various base-10 groupings. Thus, place value blocks may allow direct perception of base-10 relations that can be internalized as a conceptual metaphor for place value.

Of course, there is a sense in which such experiences could be seen as redundant. Mathematicians did not use special objects to discover mathematical formalisms. According to Lakoff & Nunez (2000), these ideas were grounded in the properties of familiar objects (e.g. numbers viewed as points along a path). If individual development recapitulates historical development in mathematics, then shouldn't experiences with everyday objects and actions (e.g. walking) be sufficient?

Perhaps, but children don't have thousands of years to reinvent mathematics from scratch. So we might think about concrete models as the 'fast track' to metaphorizing about mathematics. Specifically, these materials may provide grounding metaphors that align better with symbolic formalisms than do everyday experiences. For example, maybe mathematicians recruited walking experience as a metaphor for numbers as points on a line. However, children might not spontaneously recruit the same metaphor because there is only limited isomorphism between walking and written or spoken numbers. If, however, teachers tape a number lines across the top of children's desks, and have them move their fingers along as they count, this may generate enough direct experience that the number line itself grounds the numerals, without explicit recourse to walking. In other words, although mathematicians may have grounded their concepts of number in walking, children may 'cut to the chase' with a grounding experience that is more constrained and tailored to the corresponding symbols.

A second possibility is that concrete models act as linking metaphors— the type that connects grounding metaphors but requires direct instruction to understand. So, the number line may not be meaningful except in reference to some other experience, like walking, with this mapping provided by the teacher. Concrete models are, after all, symbols themselves. Though they are objects, they lack the functionality or relevance of everyday objects to commonplace tasks. For example, to carry a group of cookies into the other room and serve them, you need something like a plate or serving tray. Such an arrangement provides a metaphor for bounded collections of individuals, but it arises directly from the function of serving food.

In contrast, math manipulatives do not serve a function in everyday life. They are not part of common scripts. This may mean that children's interactions with them will be underconstrained because these objects serve no meaningful purpose (outside of school math instruction). Alternatively, concrete models may have inherent meaning as objects, or children may bring meaning to them by pretending they are everyday objects. For example, children could view the number lines on their desks as decorations or use them as straight-edges for drawing. But there is reason to think that this interferes with learning because such interpretations do not correspond to the objects' intended meaning as mathematical symbols (Uttal, Scudder, & Deloache 1997; Uttal et al. 2001). And when teachers provide a function, as they do for the trading chips in the Banker's Game, this is still more artificial and contrived than the function of putting cookies on a plate.

These observations suggest that at least some concrete models will not be useful unless teachers tell children explicitly what the models mean and how they relate both to concrete experience and to formal symbolism. However, once this is accomplished, concrete models could play a pivotal role as metaphors that support focused exploration in a concrete plane. For example, once children can think about the number line as a path to walk along (perhaps by 'walking' their fingers along it rather than simply sliding them), they could use this tool to practice operations, such as addition or subtraction, that would be (a) cumbersome and difficult to inspect by actually walking and (b) virtually opaque using written symbols. In the case of place value blocks, this may mean that children understand these materials only by analogy to their grounding metaphors for collections, but once this mapping has been made, they may use this new linking metaphor to interpret written place value symbols and practice related operations. Thus, even as a linking metaphor, these objects could be extremely important. A worthy question for researchers to address is which models serve this particular function (i.e. as linking metaphors) and which models are transparent enough to act as grounding metaphors.

A third potential contribution of concrete models, in terms of generating conceptual metaphors, could be simply teaching children how to create and use mathematical metaphors as a learning strategy. If metaphors are tools that foster new insights, then we need to consider not only which models are recruited, but also how children come to realize that concrete models are useful in this way. In other words, how do children discover that physical metaphors have something to do with mathematical formalisms? How do they figure out that they can generate and apply such metaphors themselves?

Instruction using math manipulatives may play a role in this regard, by modeling the process of generating and recruiting metaphors. For example, when a teacher has children practice addition and subtraction with place value blocks, they are implicitly (and sometimes explicitly) telling children that physical metaphors are related to arithmetic and can be used to bring meaning to symbolic formalisms. This means that even if children do not recruit these specific materials as conceptual metaphors, they are getting the message to seek conceptual metaphors when they are solving difficult math problems. Indeed, one explanation for individual differences in math achievement might be the degree to which different children recruit conceptual metaphors—whether actively or automatically—when they are struggling to learn new concepts. If so, then simply encouraging children to adopt this strategy could be as crucial as helping them to generate specific metaphors.

### 3.3.3 *Concrete models might offload intelligence*

Seeing cognition as embodied changes what it means to act intelligently. Intelligence no longer represents a separation from one's surroundings (i.e. a move from the perceptual to the cerebral). Instead, the learner and the situation are seamlessly united in a single cognitive system—one that seeks to relax into optimal, stable patterns whenever possible (Clark 1997; Greeno 1989; Port & van Gelder 1995). From this perspective, when learners encounter the same situation repeatedly, the probability of the same response increases (i.e. the pattern becomes more stable). And when the situation changes (i.e. it is perturbed), stable patterns are disrupted and behavioral change is possible. Importantly, this view raises the possibility that intelligence comes from *using* one's surroundings to scaffold challenging activities and achieve new insights.

Clark (1997) illustrated this idea with the example of completing a jigsaw puzzle. He pointed out that few people would start putting pieces together at random. Instead, most puzzle solvers scaffold themselves by arranging the pieces by color, or separating the edges from the interior pieces. In this way, the environment takes over some of the cognitive load. Intelligent behavior (e.g. solving the puzzle) emerges from the combination of a supportive environment and a mind with limited resources.

There are numerous examples of these 'intelligent environments'. Cooks preparing Thanksgiving dinner might post a schedule on the cabinet door, or lay pre-measured ingredients out in order. Writers can use written language to organize their thoughts and free up resources for new insights—writing one section, reading it, getting a new idea, adding it, and so forth. People can lighten the cognitive load of driving by following the car in front of them until they come to a critical intersection. These examples illustrate that what we commonly consider intelligence is not only what happens inside the brain. Instead, it is the product of the brain operating within certain environmental conditions.

From this perspective, concrete models for mathematics can be seen as features of the environment that (a) scaffold new understandings by taking over some of the cognitive load and (b) contribute to stable patterns by eliciting certain behaviors. For example, to solve the written problem '2 + 4', children must recall that the symbol '2' means two things and the symbol '4' means four things. If children's understanding of the numerals is weak, they may have trouble remembering these referents. This may mean that so many cognitive resources are taken up interpreting the numerals that there are not enough left to consider what happens when you put those two sets together. But if children represent the addends with their fingers, this could free up enough resources to allow new insights about addition to emerge, or to notice and correct errors. In this way, the child's fingers act as placeholders to offload some of the cognitive demands of addition.

A related benefit is that concrete materials provide static, external referents that can be revisited easily. In a sense, they freeze an idea in time and hold it there, at the learner's disposal. This is similar to what allows writers to use their own writing to generate new ideas. Once an idea has been put into writing, it can be inspected and analyzed in a way that is more difficult when it is purely mental and inchoate. This is because other cognitive demands make it challenging to hold a particular thought in mind indefinitely. By writing an idea down, it takes on a stability that permits deeper analysis. Similarly, when an addition problem is represented using blocks, the problem itself becomes an object that can be considered further. Children can reverse the addition process, repeat it, or just recount the solution. This may be particularly useful for complex, multi-step problems, such as subtraction with borrowing, where each step is represented by a distinct object state.

So, when teachers use concrete models to teach mathematics, they may be providing the environmental part of the intelligence equation. In a sense, they are using these objects to pull ideas out of children that might not emerge on their own. This view is quite a contrast with the alternative in which teachers provide the ideas and children, like vessels, are filled up with them. However, it is not an entirely new advance. Vygotsky wrote extensively about the ways cultural tools can scaffold young learners (e.g. Bruner 1965; Rogoff & Chavajay 1995; Vygotsky 1978). The idea of the 'prepared environment' is also a core principle in Montessori's educational approach (Hainstock 1977; Lillard 2005). What may be new here is that cognitive science can better explain *why* these approaches work—by offloading some of the burden to the environment so that the inherent limits of our memory and attention can be overcome.

This view of intelligence raises some key questions about the ultimate goal of development. In the concrete-to-abstract view, it seemed that the goal was to get by with less scaffolding. In fact, decades ago, children were prevented from using their fingers to calculate because it was believed that this 'crutch' would interfere with the development of abstract calculation. However, it now appears that the endpoint of development is not context-free thought. Instead, adults continue to rely on supportive structures in the environment. So, what develops may be the ability to generate increasingly effective scaffolds.

Research on children's addition strategies provides support for this hypothesis (Siegler & Robinson 1982). Young children typically use their fingers to solve simple calculation problems, such as '2 + 4 = 6', and they pass through a consistent series of strategies. At first, they use the 'count all' strategy, in which they raise fingers on each hand for each addend, and then count the entire set. Over time, a more efficient strategy emerges in which children represent one addend with their fingers and count it, starting with the other addend. So they might raise four fingers, but count them, '3-4-5-6'. It is faster to start with the larger number—by raising two fingers, for example, and counting them, '5-6'. Children eventually discover this and begin counting on from the larger number, whether or not the original problem is presented in that order (e.g. '2 + 4' or '4 + 2'). When memory for the basic number facts becomes automatic, children find that this is both faster and less error-prone than calculation with fingers, and switch to using that strategy most frequently. At the heart of this trend toward more efficient strategies is a trend toward increasingly efficient scaffolds, in which fingers are used in new and creative ways, and language (the ultimate scaffold) is gradually incorporated until fingers are no longer needed.

Perhaps this is the progression for all scaffolds, including concrete models for mathematics. When concepts are unfamiliar, children may need to scaffold every aspect of a problem in order to grasp it at any level. Painstakingly slow use of concrete objects may be worth the effort if it is the only way to access an idea. However, as children become more skilled with the objects and can use language or written symbols to scaffold certain aspects of a problem, they may invent faster and more efficient uses of the objects. Eventually, they may find linguistic scaffolds to be sufficient—preferable, in fact, if these tools can be used with less effort. If this is the progression, then an important question for teachers is whether children should be prodded along this path or allowed to traverse it at their own pace.

Another question concerns whether stable patterns involving concrete models are generalizable. From the embodiment perspective, children cannot learn (i.e. produce stable responses) unless there is stability in the situations themselves. Yet, if these responses are reliable only for a narrow set of environmental conditions, then their usefulness is seriously limited. Such behaviors run the risk of becoming cognitive 'backwaters'—an end unto themselves without broader implications. For this reason, Thelen (1995) argued that it is better to use multiple, overlapping contexts early in learning as well as implementing a learning sequence with successive levels of generalization. However, this raises new questions, such as how much overlap among contexts is acceptable (or optimal). And at what pace generalizations should be introduced. Unfortunately, there are no solid research-based guidelines for teachers to use in making such decisions.

### 3.3.4 *Concrete models might focus attention*

Learning requires selective attention. With nearly unlimited streams of information to process, selective attention is the gatekeeper that allows certain information in and screens out the rest. And what gets in can have profound effects on subsequent development. Smith and colleagues have shown repeatedly how improved selectivity (e.g. attention toward shape) accelerates early word learning (Landau, Smith, & Jones 1988; Samuelson & Smith 1998; Smith, Jones, & Landau 1996). Other studies have revealed the importance of joint attention in the social construction of knowledge (e.g. Baldwin 1991). And

Yu and Ballard (Chapter 10 below) demonstrated that a computer learning algorithm can pick out words from the speech stream and assign meaning to them by attending to where a 'teacher' is pointing while reading a picture book. Indeed, the establishment of joint attention is so fundamental to the enterprise of teaching that it is hard to imagine how instruction of any kind could take place without it.

This suggests a fourth and final role for concrete models for mathematics. These materials are designed to isolate mathematically relevant patterns in a way that everyday objects do not. By their very structure, they direct attention toward certain relations. This means that a likely benefit of working with math manipulatives is simply having one's attention focused on the relevant information. To illustrate, consider learning about equivalent fractions. It is certainly possible that children could discover 1/2 is equal to 4/8 by eating pizza. But these relations may not capture children's attention when there are so many other aspects competing for it, such as the way the pizza smells, who's going to eat the last piece, and picking off the mushrooms. In contrast, when children are given plastic 'pizzas'—small featureless disks that are divided into halves and eighths—the most salient attribute may be the relative size of the pieces. Just being exposed to these materials, without further instruction, may be enough to shift children's attention toward fractional relations.

Concrete models also provide a referent for joint attention. Although a teacher could just tell students that 1/2 is the same as 4/8, this would require students to conjure up their own examples—examples that the teacher could not access easily to check for understanding. Communication is clearly facilitated by reference to an example that both parties can observe. Stop and consider how many discussions about cognitive science have relied on the manipulation of pop cans and coffee cups! Concrete models for mathematics may play the same role—giving teachers and students something to talk about (Thompson 1994). In this regard, models that are specifically designed for teaching math may be particularly useful because irrelevant and potentially distracting features have been stripped away.

We can think of this as an extended cognitive system, in the sense of offloading intelligence mentioned previously. But in this system, the teacher is included along with the learner and the supporting materials. Obviously, this system could change for many reasons (what the teacher says, what the student says, etc.). But the system could also change when either the student or the teacher manipulates the materials. Thus, teachers can use the materials to make their ideas explicit and provoke shifts in students' attention. Students can rely on the materials to scaffold new insights. And teachers can gain access to students' current level of understanding by watching how the materials are used. By viewing this system as a unified, dynamic whole, it becomes clear how concrete materials could be crucial as an attentional and conversational medium.

## 3.4  Do concrete models work?

This is the critical question for educators. Should teachers expend precious financial resources and instructional time for these materials? Are they worth it or not? For good reason, this question has guided most research on concrete models for mathematics. However, the answer has been anything but clear-cut. Instead, this seemingly straightforward question has yielded an assortment of conflicting opinions, ranging from enthusiastic endorsement to lukewarm disappointment and downright skepticism (Ball 1992; Fennema 1972; Friedman 1978; Fuson & Briars 1990; Goldstone & Sakamoto 2003; Goldstone & Son 2005; Kaminski et al. 2005; 2006a; 2006b; Moyer 2001; Peterson, Mercer, & O'Shea 1988; Resnick & Omanson 1987; Sowell 1989; Suydam & Higgins 1977; Uttal et al. 1999; Uttal et al. 2001; Wearne & Hiebert 1988). And even when the results are clearly supportive of concrete models, there are so many competing variables that it is unclear *why* they worked (Ginsburg & Golbeck 2004).

Maybe the problem lies, not with the differing results, but with the question itself. As the preceding review demonstrates, there are many ways concrete materials might help, because they could engage one of several cognitive mechanisms. And the situation is further complicated by the fact that these mechanisms are not mutually exclusive. Instead, each contributes something different, but essential, to the learning process. So, concrete models implemented a particular way could activate one mechanism, many mechanisms, or no mechanisms at all. This suggests that a better question might be, 'Do these materials used in this particular way activate this particular mechanism in this particular learner?'

To illustrate, consider computer-based math manipulatives. These tools allow children to reposition, separate, and combine pictures that represent concrete models, such as place value blocks, instead of handling the actual objects themselves. For example, Figure 3.2 presents two addends in a multi-digit addition problem represented in Thompson's (1992) 'Blocks Microworld'. To add these quantities, children press the 'combine' button to make one large pile. To express their solution in canonical form (i.e. in place value terms rather than a jumble of tens and ones, etc.), they can place blocks side by side and 'glue' them together. All the while, numerical representations of both addends, as well as the solution, are presented alongside the blocks.

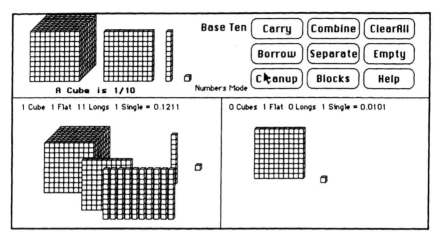

FIGURE 3.2. Screenshot from Blockworlds (Thompson 1992)

These representations include written numerals as well as expanded notation (e.g. '245' vs. '200 + 40 + 5'). The written representations are continually updated as children manipulate the blocks so that the relation between written and block representations is closely tied throughout the problem-solving process.

Thompson (1992) reported mixed results for fourth graders who had been taught using the Blocks Microworld program versus those taught with actual blocks. Neither group showed significant improvement from pretest to post-test on whole-number calculation problems. Children in the physical blocks group performed somewhat better with decimal computation, whereas children taught with Blocks Microworld exhibited better understanding of decimal ordering and equivalence and more flexible use of written symbols. So do concrete models work, or don't they? It appears that concrete objects might work and Blocks Microworld might work, but it depends on characteristics of the learners, the problems, and the outcome measures.

This conclusion will be familiar to scholars in math education. Many research papers on the effectiveness of math manipulatives end with a list of factors that might explain discrepant results (e.g. Resnick & Omanson 1987; Thompson 1992). And even those who advocate the use of math manipulatives caution that these materials are not intrinsically effective. Instead, their effectiveness depends on the way they are implemented (Baroody 1989; 1990; Simon 1995; Post 1988). The problem is that simply identifying sources of variation is not enough. Researchers need to figure out why these factors matter and how they interact.

With this in mind, let us reconsider Thompson's (1992) results in light of the learning mechanisms discussed previously. First, recall that none of the children improved their whole number calculation scores from pretest to posttest. Thompson's interpretation was that children's symbolic procedures for whole-number calculation were already so entrenched that children were impervious to input from either the concrete or computerized block manipulations. This interpretation is consistent with the mechanisms of perception-action learning. That is, entrenched procedural knowledge could be construed as a very stable pattern that arises in response to a particular situation (e.g. written whole-number problems). However, this reconstrual is more than a difference in semantics. These two views have very different implications for practical applications and future research.

The view that children's whole-number calculation procedures were entrenched implies inevitability—a static state in which concrete models no longer have an impact. This seems to argue against using this instruction to improve whole-number calculation in such children. Furthermore, though this interpretation suggests that further research with younger children (i.e. those with less exposure to whole-number calculation) would reveal an effect of concrete manipulatives, it provides no reason to seek such effects in fourth graders.

In contrast, the view that children's performance reflected a stable but context-dependent pattern implies fluidity—the idea that even highly stable patterns can be destabilized and changed. On this view, enough exposure to concrete models should create other stable patterns that would eventually be strong enough to compete with the existing symbolic procedures. The precise amount of exposure needed should vary depending on how stable each child's existing responses already were, but it should be possible no matter how 'stuck' a pattern may seem. This illustrates how exposure and learning history would interact to produce different behavioral patterns at different levels of each factor. Another implication of this view is that there should be other ways to perturb the stable pattern, such as changing the testing context. Perhaps if children were asked to respond in a different way (e.g. not completing a worksheet of written problems), they would reveal new learning from concrete models after less exposure because a novel context would be less likely to activate the stable, whole-number calculation pattern. Here, task interacts with time and learning history to produce still more possible outcomes. The main point is that research on these questions could make a much greater contribution if it focused on the way these factors operate as a system.

As a further illustration, consider Thompson's (1992) finding that children trained in the Blocks Microworld exhibited better place value comprehension than children trained with actual objects. From the perception-action perspective, this finding seems surprising. After all, if cognition is built from movements in a physical environment, then direct experience acting upon concrete objects *should* be critical. Furthermore, the idea that conceptual metaphors arise from bodily experiences, such as dressing and undressing, seems to favor contact with actual objects. In short, if children learn through actions and direct experiences, how is this finding possible?

Thompson (1992) offered two explanations. One was that the microworld program better supported learning about written symbols by providing effortless overlap between written notation and physical representations. Recall that whenever the virtual blocks were manipulated, the computer provided a written representation, in regular and expanded notation, for the represented quantities. The idea was that having this notation continuously available helped children map between the two representations. The second explanation was that the computer microworld constrained the possible actions to mathematically relevant ones, whereas direct contact permitted other actions that may prove distracting.

The problem is that these explanations only go so far because, though sensible given the data, they do not specify the cognitive mechanisms involved. Nor, for that matter, is there sufficient evidence to conclude that these explanations are basically correct. For example, the first explanation assumes that the presence of written numerals in the computer displays is what matters. But to conclude this, we would need to know that children learn less from an identical computer program that omits written notation. In other words, with so many differences between the concrete and computer instruction, it is unclear which particular difference matters without isolating and testing each one separately.

If it were clear that the written notation helped, this effect could be based on several different mechanisms that each carry different implications. One would be offloading intelligence. By generating written representations automatically, the computer frees the learner from the demands of recalling and writing the correct numerals. However, the availability of written representations also provides many examples of symbols juxtaposed with pictures of objects. So, a second mechanism could be massive exposure to symbol-to-referent mappings, leading to a stable pattern of response given correlations between these two information streams. Yet a third possibility is that the continuous presence of written numerals, featured prominently in the computer display, may direct attention to them. Which mechanism is it? Could it be all three?

To find out, researchers might test whether children perform as well with real blocks if a tutor or partner writes down all the numerals for them, or if they are given worksheets that scaffold the recording process, thus alleviating cognitive load. Or researchers might vary cognitive load independent of the mapping process by changing the problem types or adding irrelevant demands, like sorting the blocks by color during training. To isolate the effect of mapping experience, researchers could compare conditions where children are taught with real objects but are given extra mapping practice or training. If similar effects can be achieved with concrete materials under these conditions, we can conclude with greater certainty that these features are what make the computer program superior. The main point is that breaking these explanations down into their cognitive components points the way toward testing them with more rigor and precision. In the end, it is not enough to know whether one instructional approach is better than another. What begs to be discovered is *why*.

## 3.5  Instructional issues revisited

Earlier in this chapter, I outlined some of the choices facing teachers who incorporate concrete materials into their mathematics instruction. One consequence of viewing this enterprise in terms of cognitive mechanisms is that many of these apparent 'either/or' propositions transform into delicate balancing acts. In this section, I will illustrate this point by considering just a few of these. Recall that one fundamental instructional decision is whether to use one concrete representation or many. Different theorists have developed excellent arguments on both sides of this question. Some have argued that one material is best because it fosters deep expertise and allows children to see subtle interrelationships across operations (Stevenson & Stigler 1992). Others have claimed that children can only identify abstract patterns by analyzing the commonalities among disparate examples (e.g. Dienes 1960). However, modern approaches to cognitive science suggest that both approaches are necessary. What is at issue is timing.

If we think of learning as increasingly stable responses to the constraints of a particular situation (e.g. Thelen 1995), the situation must be stable for the response to be stable. This means that children will need enough time with a particular material to achieve a degree of stability. However, too much stability in a situation could lead to rigidity in the response. That is, if the features of a situation are completely consistent across instances, the behavior might become encapsulated, elicited only in one narrow context. To combat this problem, Thelen recommended varying the situations. But the key here—and what sets this view apart from previous conceptualizations—is that the amount of variation should be just enough: enough variation to stop the behavior from getting stuck but not so much that the target responses patterns are no longer elicited. This means that the decision of one versus many models is not an all-or-nothing proposition. Instead, it is a matter of finding the critical balancing point—a point that will vary for different children at different points in learning. This explains, with greater clarity, why concrete manipulatives may have limited benefit unless they are used in individualized or small group instruction (Post 1988).

One instructional approach that may help teachers find this balance is the gradual introduction of new but overlapping problem situations. This is the cornerstone of Montessori's approach to sensorial math instruction (Lillard 2005), and is generally consistent with other instructional recommendations (e.g. Baroody 1989; Bruner 1960; Miller & Mercer 1993; Peck & Jencks 1987). Its effectiveness also has been demonstrated in various training experiments. For example, Kotovsky & Gentner (1996) successfully trained 4-year-olds to recognize abstract relations among schematic pictures, such as two sets of shapes that both increased in size monotonically, with a series of progressive alignment trials. Specifically, children were taught to match nearly identical object sets that increased in size before attempting comparisons with fewer surface features in common. This learning effect seems consistent with the idea that stable patterns must be fostered in a single problem situation before moving on to new situations. It also illustrates that new understandings first emerge in maximally supportive situations (e.g. identity matches), but can be generalized if the move toward abstraction is progressive.

In a similar vein, Goldstone & Son (2005) found that 'concreteness fading' was the most effective way to teach undergraduates about complex adaptive systems. Students were first allowed to manipulate elements in a computer program that were detailed and concrete, such as ants foraging in different food patches. However, the computer displays gradually became more and more schematic until they consisted of nothing more than dots. Students trained with these 'fading' displays demonstrated better understanding at post-test than those who received the same amount of training with either concrete or abstract displays. Perhaps this training worked because students were allowed to develop stable patterns in a maximally supportive context, but were led away from it in a gradual way that did not sever the connection between one situation and the next.

Another decision facing teachers concerns how much to guide or constrain manipulative activities. On one extreme, children would be allowed to explore concrete materials, unconstrained, for long periods of time. On the other, teachers would direct manipulative activities, step by step, while children follow along. This decision has practical implications. Extended periods of free exploration require time and money—probably more time and money than most schools can provide. Yet there is good reason to think that children need full access to these materials with no constraints as well as guided activities with many constraints.

This decision boils down to a tension between two different cognitive mechanisms. One is learning through direct experience and action. If cognition is built from perception and action in a physical environment, then direct experience acting upon concrete objects would be critical. Furthermore, the idea that conceptual metaphors arise from bodily experiences, such as dressing and undressing, favors contact with actual objects, at least early in development. On these accounts, stable action patterns and bodily experiences are the stuff of subsequent symbol manipulation and conceptual abstraction, suggesting that this type of learning should not be short-changed.

Still, stable action patterns are a far cry from the written-calculation algorithms they are meant to illuminate. To understand what written symbols represent, activities with greater constraint and focus may be needed, such as performing specific calculation algorithms with blocks or recording every step of a calculation problem with written symbols. Herein lies the tension, because the need to find stable patterns through direct exploration competes with the need for attentional focus. When tasks are highly constrained and prescribed, attention is optimally focused. But when the teacher provides too much structure, or the wrong kind of structure, the result could be fragile patterns that are activated only under optimal conditions.

Maria Montessori addressed this tension in an interesting way. On the basis of careful observation and experimentation, she developed a variety of clever materials for teaching mathematics to young children. But in addition to the materials themselves, Montessori also developed careful protocols for children's use of them. These protocols were aimed at fostering concentration—an intense, task-oriented focus—that is the hallmark of Montessori classrooms (Hainstock 1997; Lillard 2005). What is relevant for our purposes is that 'concentration' in the Montessori sense encompasses both directed attention and embodied learning.

Concentration, in Montessori's sense, encompasses directed attention because it is promoted by constraining children's activity. In short, there is a right and a wrong way to interact with each manipulative. For example, the Pink Tower consists of cubes that increase in size from $1\,cm^3$ to $10\,cm^3$. There are many possible activities that such materials could engage, but children are not allowed to do whatever they want with them. Instead, they are shown how to carry the cubes to a rug and construct a tower in a particular way. Each of these constraints is in place for a reason. Carrying the cubes to the rug one by one allows children to notice differences in size and weight, for example (Lillard 2005). The main point is that Montessori believed children got the most out of concrete materials when they carried out actions specified by the teacher. Modeling these actions and ensuring children's compliance is one way in which teachers foster concentration.

Concentration is embodied because it is also promoted without direct intervention from the teacher. Montessori education is built on the idea of the prepared environment (Hainstock 1997; Lillard 2005). The way teachers place the furniture affects the way children move around the room. The rotating selection of activities puts a limit on the number or kind of materials children can use. And the materials themselves, by virtue of being self-correcting, tend to engage children appropriately without teacher intervention.

In fact, another key aspect of fostering concentration is that children are allowed to work uninterrupted, for hours every day. They choose what activities to complete and they decide how long to complete them. This freedom can lead to rather protracted sessions. In one famous anecdote, a girl repeated the Wooden Cylinders work 44 times without stopping. She was in such a deep state of concentration that when Dr Montessori lifted the child's chair, with the child in it, the child continued to complete the work in her lap without even noticing (Montessori 1917; 1965). This respect for concentration can be seen as respect for something else—the learner's need to consolidate perceptions and actions into stable patterns of behavior. Thus, we see both perception-action learning and explicit attempts to control attentional focus united in the service of promoting concentration. This may be a useful model for others seeking to resolve this tension.

Obviously, this section has not addressed every decision facing teachers who use concrete materials. However, I hope it has illustrated how these practical problems have theoretical implications, and how we might view these decisions differently by shifting to a framework that emphasizes the underlying cognitive mechanisms involved.

## 3.6  Conclusions

The issues involved in teaching with concrete materials may seem black and white. Do these materials work or don't they? Should we teach with one material or many? Although there may not be sufficient data to answer every question, at least the questions themselves seem straightforward. But when we dig deeper into the learning mechanisms these materials might tap, such delineations become fuzzy. Instead of dichotomies, we find continuua. Instead of simple answers, we must qualify each result in terms of a range of controlling variables. In fact, when the problem is viewed in its entirety, the obvious questions no longer seem quite right.

Some teachers manage to navigate this bramble with great success. Simon (1995) described their decision-making processes as something akin to theory-testing, where teachers plan their lessons based on hypotheses about student learning. They make decisions about where to come down on a particular dichotomy based on mathematical expertise, knowledge of their current students' abilities, and a strong intuitive sense of development based on theory, research, and, above all, experience. They constantly adjust their course as they observe the effects of one approach or another. They are literally feeling their way through.

Researchers can and should do more to inform these decisions. First, by pinning down the relevant learning mechanisms, teachers can have a clearer understanding of why one material works better than another. I have outlined several possible mechanisms in this chapter, but there may be more. In any case, direct empirical evidence of these mechanisms in children's learning about mathematics is needed. Researchers also could help by developing ways to measure learning vis a vis these mechanisms. For example, the issue of exposure time has been raised repeatedly in this chapter. It would be helpful to provide some way for teachers to know when children have had enough experience with one material and are developmentally ready to move on to something new. Such signposts would not only lighten the load for successful teachers, but might also help teachers with less intuition make better informed decisions. Finally, it seems that the time has come to move on to questions that address the multifactorial nature of this learning problem in a systematic way. This could be achieved in many ways, but a good start might be to gather more information on learner differences and use these as a framework for comparing instructional approaches.

# 4

# Time, Motion, and Meaning: The Experiential Basis of Abstract Thought

MICHAEL RAMSCAR, TEENIE MATLOCK, AND LERA BORODITSKY

In our everyday language, we often talk about things we can neither see nor touch. Whether musing on the passage of time, speculating on the motives of others, or discussing the behavior of subatomic particles, people's endeavors constantly require them to conceptualize and describe things that they cannot directly perceive or manipulate. This raises a question: how are we able to acquire and organize knowledge about things in the world to which we have no direct access in the first place? One answer to this conundrum is to suppose that abstract domains may be understood through analogical extensions from richer, more experience-based domains (Boroditsky & Ramscar 2002; Boroditsky 2000; Clark 1973; Gibbs 1994; Lakoff & Johnson 1980a). Supporting evidence for this proposal can be seen in the way people talk about concrete and abstract domains. Everyday language is replete with both literal and metaphorical language that follows this broad pattern. Take, for instance, motion language. In its literal uses, it is descriptive of paths and trajectories of objects, as in 'Bus 41 *goes* across town', 'A deer *ran* down the trail', and 'The boys *raced* up the stairs'. In its metaphoric uses, which are pervasive in everyday speech, motion language is descriptive of emotions, thought, time, and other abstract domains, as in 'He *runs* hot and cold', 'My thoughts were *racing*', and 'Spring break *came* late'. Similarly, representational structure from the domain of object motion appear to be borrowed to organize our ideas about space, including static scenes, as in 'The trail *goes* through town', 'The fence *follows* the river', or 'The tattoo *runs* down his back'.

The hypothesis that the structure of abstract knowledge is experience-based can be formulated in several strengths. A strong 'embodied' formulation might be that knowledge of abstract domains is tied directly to the body such that abstract notions are understood directly through image schemas and motor schemas (Lakoff & Johnson 1999). A milder view might be that abstract knowledge is based on representations of more experience-based domains that are functionally separable from those directly involved in sensorimotor experience.

In this chapter we review a number of studies that indicate that people's understanding of the abstract domain of time supervenes on their more concrete knowledge and experience of the motion of objects in space. First, we show that people's representations of time are so intimately dependent on real motion through space that when people engage in particular types of thinking about things moving through space (e.g. embarking on a train journey, or urging on a horse in a race), they unwittingly also change how they think about time. Second, and contrary to the very strong embodied view, we show that abstract thinking is more closely linked to *representations* of more experience-based domains than it is to the physical experience itself.

Following from this, we explore the extent to which basing abstract knowledge on more concrete knowledge is a pervasive aspect of cognition, examining whether thought about one abstract, non-literal type of motion called 'fictive motion' can influence the way people reason about another, more abstract concept, time. Once again, our results suggest that metaphorical knowledge about motion appears to utilize the same structures that are used in understanding literal motion. Further, it appears that the activation of these 'literal' aspects of fictive motion serve to influence temporal reasoning. The results we describe provide striking evidence of the intimate connections between our abstract ideas and the more concrete, experiential knowledge on which they are based.

## 4.1  Representations of space and time

Suppose you are told that next Wednesday's meeting has been moved forward two days. What day is the meeting now that it has been rescheduled? The answer to this question depends on how you choose to think about time. If you think of yourself as moving forward through time (the ego-moving perspective), then moving a meeting 'forward' is moving it further in your direction of motion—that is, from Wednesday to Friday. If, on the other hand, you think of time as coming toward you (the time-moving perspective), then moving a meeting 'forward' is moving it closer to you—that is, from Wednesday to Monday (Boroditsky 2000; McGlone & Harding 1998; McTaggart 1908). In a neutral context, people are about equally likely to think of themselves as moving through time as they are to think of time as coming toward them, and so are equally likely to say that the meeting has been moved to Friday (the ego-moving answer) as to Monday (the time-moving answer) (Boroditsky 2000; McGlone & Harding 1998).

But where do these representations of time come from? Is thinking about moving through time based on our more concrete experiences of moving through space? If representations of time are indeed built on representations of space, then activating different types of spatial representation should influence how people think about time.

To investigate the relationship between spatial experience and people's thinking about time, Boroditsky & Ramscar (2002) asked 333 visitors to San Francisco International Airport the ambiguous question about Wednesday's meeting described above. After the participants answered, they were asked whether they were waiting for someone to arrive, waiting to depart, or had just flown in. Two questions were of interest: (1) whether a recent, lengthy experience of moving through space would make people more likely to take the ego-moving perspective on time (think of themselves as moving through time as opposed to thinking of time as coming toward them), and (2) whether this effect required the actual experience of motion, or if just thinking about motion was enough.

As shown in Figure 4.1, people who had just flown in were much more likely to take the ego-moving perspective (think of themselves as moving through time and answer 'Friday') (76%) than people who were just waiting

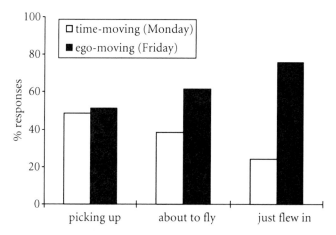

FIGURE 4.1. Responses of 333 people queried at the airport. People who had just flown in were most likely to produce an ego-moving response (say that next Wednesday's meeting had been 'moved forward' to Friday).

for someone to arrive (51%). Further, even people who had not yet flown, but were only waiting to depart were already more likely to think of themselves as moving through time (62%) (Boroditsky & Ramscar 2002). This set of findings suggests that (1) people's ideas about time are indeed intimately related to their representations of space, and (2) just thinking about spatial motion is sufficient to change one's thinking about time. But this also raises an interesting question: why were people who had just flown in more likely to take an ego-moving perspective than people who were only about to depart? Was it because they had spent more time actually moving through space, or was it just because they had had more time to think about it?

To investigate this question, Boroditsky & Ramscar (2002) posed the ambiguous question about Wednesday's meeting to 219 patrons of CalTrain (a commuter train line connecting San Francisco and San Jose). Of these, 101 were people waiting for the train, and 118 were passengers actually on the train. All of them were seated at the time that they were approached by the experimenter. After participants answered the question, they were asked how long they had been waiting for (or been on) the train, and how much further they had to go.

It turned out that both people waiting for the train and people actually riding on the train were more likely to take the ego-moving perspective (63%) than the time-moving perspective (37%). Interestingly, the data from people waiting for the train looked no different from those of people actually on the

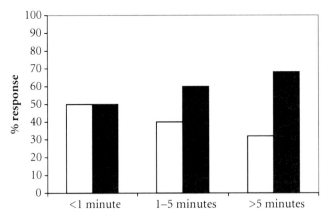

FIGURE 4.2. Responses of 101 people waiting for the train plotted by time spent waiting. The more time people had to anticipate their journey, the more likely they became to adopt the ego-moving perspective on time (say that next Wednesday's meeting has been 'moved forward' to Friday).

train (61% and 64% ego-moving response respectively), suggesting that it is not the experience of spatial motion *per se*, but rather thinking about spatial motion that underlies our representation of time.

Boroditsky & Ramscar (2002) then examined people's responses on the basis of how long they had been waiting for the train (see Figure 4.2). The longer people sat around thinking about their journey, the more likely they were to take the ego-moving perspective for time. People who had waited less than a minute were equally as likely to think of themselves as moving through time as they were to think of time as coming toward them. People who had had five minutes of anticipating their journey were much more likely to take the ego-moving perspective on time (68%) when compared to people waiting less than a minute (50%).

Finally, the responses of people on the train were analyzed on the basis of whether they had answered the ambiguous time question at the beginning, middle, or end of their journey. The conjecture was that people should be most involved in thinking about their journey when they had just boarded the

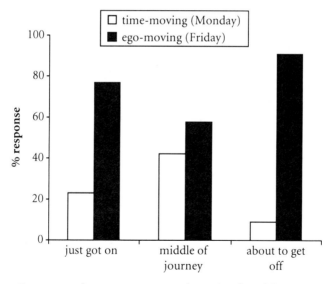

FIGURE 4.3. Responses of 118 passengers on the train plotted by point in journey. People became much more likely to adopt the ego-moving perspective for time (say that next Wednesday's meeting has been 'moved forward' to Friday) when they were most engaged in thinking about their spatial journey (at the beginnings and ends of the trip). In the middle of their journey, people were about equally likely to adopt the ego-moving perspective (say the meeting has been 'moved forward' to Friday) as the time-moving perspective (say the meeting has been 'moved forward' to Monday).

train, or when they were getting close to their destination. In the middle of their journey, people tend to relax, read, talk loudly on cellphones, and otherwise mentally disengage from being on the train.

It turned out that people's biases for thinking about time perfectly mimicked their patterns of engaging in and disengaging from spatial-motion thinking (see Figure 4.3). Within five minutes of getting on the train, people were very likely to be taking the ego-moving perspective on time (78%) when compared to people in the middle of their journey, who showed no significant ego-moving bias (54% ego-moving). However, people were likely to readopt the ego-moving perspective when they were within ten minutes of arriving at their destination (80% showed an ego-moving bias). Once again, it appears that people's thinking about time was affected by their engaging in thinking about spatial motion, and not simply by their experience of motion itself. Although all three groups of passengers were having the same physical experience (simply sitting on the train), the two groups that were most likely to be involved in thinking about their journey showed the most change in their thinking about time (Boroditsky & Ramscar 2002).

So far, we have only looked at people who themselves were moving or planning to move. Could thinking about spatial motion have a similar effect even when people are not planning any of their own motion? To investigate this question, we asked the 'Next Wednesday's meeting...' question of 53 visitors to the Bay Meadows racetrack. We predicted that the more involved people were in the forward motion of the racehorses, the more likely they would also be to take the ego-moving perspective on time (and say that the meeting has been moved to Friday). After asking people the question about next Wednesday's meeting, we also asked them how many races they had watched that day and how many races they had bet on. Both indices turned out to be good predictors of people's answers to the 'Next Wednesday's meeting...' question. As shown in Figure 4.4, people who had not bet on any races were as likely to think of themselves as moving through time (50% said 'Friday'), as they were to think of time as coming toward them (50% said 'Monday'). In contrast, people who had bet on three races or more were three times more likely to think of themselves as moving through time (76% said 'Friday') than they were to think of time as coming toward them (24% said 'Monday') when compared to people who had not bet on any races (50%). It appears that simply thinking about forward motion (without planning to actually go anywhere) is enough to change people's thinking about time.

The experiments described so far indicate that people's thinking about spatial motion is a good predictor of their thinking about time, and that actual physical motion may not necessarily influence co-occurrent thinking about

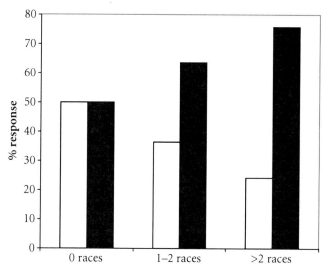

FIGURE 4.4. Responses of 53 visitors to the racetrack plotted by number of races bet on. People who had bet on more races (and so were more involved in the forward motions of the racehorses) also became much more likely to adopt the ego-moving perspective for time (say that next Wednesday's meeting has been 'moved forward' to Friday).

time. This then raises the question of whether actual motion is even *sufficient* to influence people's thinking about time, even in the absence of involved spatial thinking.

To address this question, we set up a 25-ft track outside the Stanford University Bookstore and invited students to participate in an 'office chair rodeo'. Half of the participants were asked to ride an office chair from one end of the track to the other (the ego-moving prime), and half were asked to rope the chair in from the opposite end of the track (the time-moving prime) (see Figure 4.5 for an illustration of the basic experimental set-up). The track was marked out in the asphalt using colored masking tape, with one end of the track marked in red and the other in yellow. Fifty Stanford undergraduates participated in the study in exchange for lollipops. The verbal instructions were the same in both conditions. Participants riding the chair sat in an office chair at one end of the track and were asked to 'maneuver the chair to the red/yellow line' (whichever was at the opposite end of the track). Participants roping the chair were given a rope that was connected to the office chair at the opposite end of the track and were likewise instructed to 'maneuver the chair to the red/yellow line' (whichever was where the participant was standing).

A                  Riding the chair (ego-moving prime)

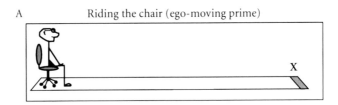

B            Roping the chair (time-moving prime)

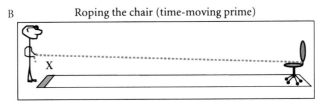

FIGURE 4.5A. The ego-moving priming materials used in the 'imagined motion' study. Participants were given the following instructions: 'Imagine you are the person in the picture. Notice there is a chair on wheels, and a track. You are sitting in the chair. While sitting in the chair, imagine how you would maneuver the chair to the X. Draw an arrow indicating the path of motion.'

FIGURE 4.5B. In this condition participants were asked to, 'Imagine you are the person in the picture. Notice there is a chair on wheels, and a track. You are holding a rope attached to the chair. With the rope, imagine how you would maneuver the chair to the X. Draw an arrow indicating the path of motion.'

Immediately after the participant completed the motion task (either riding or roping the chair), they were asked the question about next Wednesday's meeting. We found that performing these spatial motion tasks had no effect on subjects' thinking about time. People riding the chair (actually moving through space) were as likely to think of themselves as moving through time (56% said the meeting would be on Friday) as were people roping the chair (actually making an object move toward them) (52% said the meeting would be on Friday).

In contrast, we found that asking people to *think* about this task affected the way they subsequently thought about time. We asked 239 Stanford undergraduates to fill out a one-page questionnaire that contained a spatial prime followed by the ambiguous 'Next Wednesday's meeting …' question described above. The spatial primes (shown in Figure 4.5) were designed to get people to think about themselves moving through space in an office chair (see Figure 4.5a) or about making an office chair come toward them through space (see Figure 4.5b). In both cases, participants were asked to imagine how they would 'maneuver the chair to the X', and to 'draw an arrow indicating the path of motion'. The left-right orientation of the diagrams was counterbalanced across subjects. After our subjects completed the spatial prime, they were asked the ambiguous 'Next Wednesday's meeting …' question.

Our results indicated that in contrast to *actually* moving, *imagining* themselves as moving through space, or *imagining* things coming toward them, did cause our participants to think differently about time. Subjects primed to think of objects coming toward them through space were more likely to think of time as coming toward them (67% said Wednesday's meeting had moved to Monday), than they were to think of themselves as moving through time (only 33% said the meeting had moved to Friday). Subjects primed to think of themselves as moving through space showed the opposite pattern (only 43% said Monday, and 57% said Friday) (Boroditsky & Ramscar 2002).

It appears that just moving through space, without thinking much about it, is not sufficient to influence people's thinking about time. In contrast, imaging the self-same experience does influence people's thinking about time. This finding is especially striking when taken in conjunction with previous evidence that just thinking about spatial motion (in the absence of any actual motion) is enough to influence people's thinking about time (Boroditsky 2000).

Taken together, the studies described so far demonstrate an intimate relationship between abstract thinking and more experience-based forms of knowledge. People's thinking about time is closely linked to their spatial thinking. When people engage in particular types of spatial thinking (e.g. thinking about their journey on a train, or urging on a horse in a race), they also unwittingly and dramatically change how they think about time. Further, and contrary to the very strong embodied view, it appears that this kind of abstract thinking is built on representations of more experience-based domains that are functionally separable from those directly involved in sensorimotor experience itself (see also Boroditsky & Ramscar 2002).

## 4.2 Fictive representations of space and their influence on the construction of time

So far we have seen that *thinking* about objects moving through space can influence the way people conceptualize the 'motion' of time. That is, thinking about concrete motion seems to have affected the way people subsequently thought about a more abstract domain that borrows structure from that more concrete parent domain. We now turn to the relationship between fictive motion and thinking about time.

Fictive motion sentences (e.g. 'The tattoo runs along his spine' or 'The road goes along the coast') are somewhat paradoxical because they include a motion verb ('run', 'go') and physical scene ('spine', 'coast'), but they describe no physical movement or state change (Matlock 2004; Talmy 1996). However, in language after language they systematically derive from literal uses, which *do* describe physical movement (e.g. 'Bus 41 <u>goes</u> across town'; Radden 1996; Sweetser 1990; Miller & Johnson-Laird 1976). The ubiquity and diachronic regularity of fictive-motion language provides further support for the idea that people recruit experiential concepts acquired from the physical world to make sense of more abstract domains. Further, it allows us to pose and explore an intriguing question: Can the borrowed structures from real motion understanding—used to flesh out our understanding of spatial relations in fictive motion—be used to influence similar borrowed structures in the temporal domain, so as to affect people's conceptions of time?

Does fictive motion involve the same conceptual structures as real motion? If so, manipulating people's thinking about fictive motion should also influence their temporal thinking. To examine this, in a series of apparently unrelated questionnaire tasks we asked 142 Stanford University students to: (a) read either a fictive motion sentence (hereafter, FM-sentence) (e.g. 'The road runs along the coast') or a comparable no-motion sentence (hereafter, NM-sentence) (e.g. 'The road is next to the coast'), (b) sketch the spatial scene described by the sentence (the drawing task made sure participants paid attention to and understood the sentence), and (c) answer the ambiguous temporal question 'Next Wednesday's meeting has been moved forward two days. What day is the meeting now that it has been rescheduled?' We wanted to see whether sentence type would influence response (Monday versus Friday). Critically, if participants mentally simulate scanning along a path (see Matlock 2004; Talmy 1996; 2000), this would be congruent with an ego-moving actual motion perspective (Boroditsky 2000); if they are simulating motion with fictive motion, it ought to encourage them to think of themselves (or some other agent—see Boroditsky & Ramscar 2002) 'moving' through time as they scan motion, prompting a Friday response.

We found that the fictive motion primes did influence our participants' responses to the ambiguous temporal question. FM-sentences led to more Fridays than Mondays, but NM-sentences showed no difference. Of the participants primed with fictive motion, 70% went on to say the meeting would be Friday, and 30% said Monday. In contrast, 51% of those primed with no-motion went on to say Friday, and 49% said Monday—a close but statistically reliable difference (Matlock, Ramscar, & Boroditsky 2005).

These results indicate that thought about fictive motion does indeed influence thought about time. When people process fictive motion, it appears that they apply the same motion perspective to their thinking about time as when

(1) No motion: *The bike path is next to the creek*

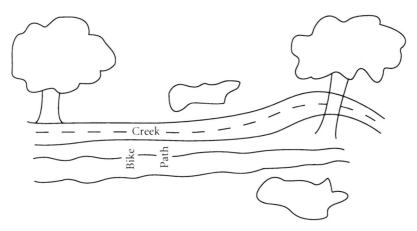

(2) **Fictive motion:** *The bike path runs alongside the creek*

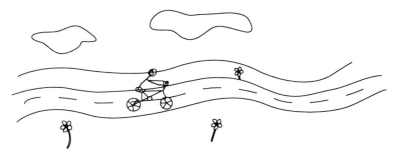

FIGURE 4.6. Examples of drawings with no motion sentences and fictive motion sentences
(a) No motion: *The bike path is next to the creek*
(b) Fictive motion: *The bike path runs alongside the creek*

they process actual motion. In this case, they appear to subjectively scan a path, and this accordingly activates an ego-moving schema, which in turn produces a Friday answer. When they think about a comparable spatial description without fictive motion and which does not relate to a particular motion schema, their temporal thinking is unaffected, and hence in answering an ambiguous question about time, their responses are at chance.

This raises the question of what it is about fictive motion that affects temporal thought. If fictive motion really is activating some abstract representation of concrete motion, then the effects we observed above might vary according to the amount of 'motion' in a given fictive motion prime. That is, we might expect the fictive motion effect to be more robust with a 'longer' fictive path than with a 'shorter' fictive path (see Figure 4.6).

To examine this, we examined 124 Stanford students using a procedure similar to the one described above. In this experiment, however, we varied the length of the path of the fictive motion by asking our participants to read one of the following sentences: 'Four pine trees run along the driveway, Eight pine trees run along the edge of the driveway, Twenty pine trees run along the edge of the driveway, Over eighty pine trees run along the edge of the driveway'. We reasoned that if people activate conceptual structure about motion while thinking about fictive motion, then we should expect more (e.g. longer) motion simulation when people can conceptualize more points along the scan path. Further, given the finite resources available to people in working memory, we also predicted that (as the old saying about not seeing the wood for the trees suggests) if people had an indeterminately high number of trees to individuate as scan points in conceptualizing the over-80-tree FM-sentence, such that their representational capacities for individual trees were swamped, they might tend to conceive of 'many trees' as a mass entity. In this case, this might function as a poor prime because its representation would possess few scan points.

Since more scanning in simulation should be more likely to activate an ego-moving perspective when thinking about time, we expected that we would see more Fridays than Mondays in response to the question as the number of scan points increased from 4 to 8 to 20, but a drop in this effect as the number of trees increased to over 80.

This is what we found. As shown in Figure 4.7, there was a significant interaction between sentence type and number of pine trees. These results indicate that responses were differentially influenced by the way people had thought about fictive motion, in this case by the number of points along a path. As shown in sample drawings in Figure 4.8, 8 and 20 trees were sufficient in number (not too many, not too few) for people to build up an adequate path representation—that is, one along which people could simulate motion or visual scanning. A total of 4 trees, however, did not allow people to produce an adequate path representation, and a total of over 80 trees was too many.

In sum, people were more likely to respond 'Friday' than 'Monday' when they could simulate motion along a just-right-sized path (when they had thought about 8 trees or 20 trees running along a driveway), but there was no reliable difference when people had thought about only 4 trees or over 80 trees. This suggests that people built a path representation upon reading a fictive motion sentence, and that this was then incorporated into the representations

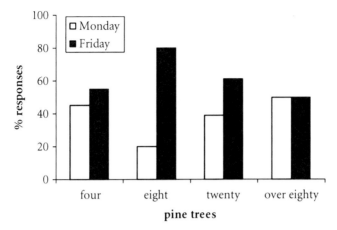

FIGURE 4.7. Responses to the ambiguous question plotted by the number of pine trees in the prompt

they used to reason about when the meeting would be held. When the number of trees was more conducive to building a representation that could be readily scanned (not too few, not too many), people were more prone to adopt an ego-moving perspective (see Matlock et al. 2005).

So far we have seen that thinking about fictive motion influences the way people think about time, but we have not ascertained whether fictive motion involves a diffuse or abstract sense of motion or a more defined sense of directed motion. To explore the extent to which fictive motion construal involves direction, an important conceptual property of motion construal (Miller & Johnson-Laird 1976), we primed 74 Stanford students with a FM-sentence about a road beginning at an unspecified location and terminating at a far-away location (New York), or a sentence that begins at the far-away location and 'moves' toward the unspecified location, to see whether people would construct a representation in which they were either the starting point or ending point of a path. If so, thinking about the road 'going' toward New York might encourage a 'Friday' response consistent with the ego-moving perspective where individuals see themselves moving through time ('Monday is ahead of me'). This is analogous to the ego-moving perspective in actual motion, where, when individuals construe themselves as moving through space, the 'front' object will be that which is furthest away. If participants thought about the road 'coming' to them, we expected a Monday response, consistent with a time-moving perspective in which the individual is seen as stationary, with events coming towards them ('Christmas is coming'). This is analogous to the

*Twenty pine trees run along the edge of the driveway*

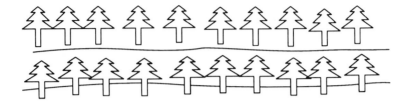

*Over eighty pine trees run along the edge of the driveway*

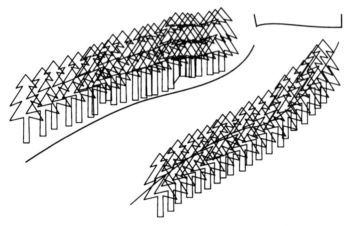

FIGURE 4.8. Examples of drawings for different numbers of trees
(a) *Twenty pine trees run along the edge of the driveway*
(b) *Over eighty pine trees run along the edge of the driveway*

object-moving perspective in actual motion, where, when individuals construe objects as moving towards themselves as moving, the 'front' object will be closest to an observer (Boroditsky 2000).

Of the participants primed with fictive motion 'towards themselves' ('The road comes all the way from New York'), 62% responded Monday and 38% Friday, and of the participants primed with fictive motion 'away from themselves' ('The road goes all the way to New York'), 33% responded Monday and 67% Friday (Matlock et al. 2005). The results indicate that people were influenced by their understanding of fictive motion. When people thought about fictive motion *going* away from themselves (Stanford), they appeared to adopt an ego-moving perspective and conceptually 'moved' while time remained stationary. In contrast, when people engaged in thought about fictive motion *coming* toward them (and their location, Stanford), they appeared to adopt a perspective whereby they remained stationary and time moved toward them. These results suggest that fictive motion involves simulating motion along a path, and that that motion can be directed.

As we noted earlier, it is far from obvious that thinking about fictive motion *should* bring about any differences whatsoever in the way people think about time, especially given that nothing actually moves in a fictive motion description. In the real world, tattoos do not move independently of the skin upon which they are inked, and bookcases do not run around rooms. The subject noun phrase referents in fictive motion sentences, such as 'tattoo' in 'The tattoo runs along his spine', are in no way actually moving. Because of this, the question of whether fictive motion involves a dynamic conceptualization has long been controversial. Talmy (2000; 1996) and Langacker (2000) have proposed that the representation underlying fictive motion sentences may be temporal, dynamic, and involve structures akin to real motion. Matlock's (2004) results provide empirical evidence to support this idea. Counter to this, however, Jackendoff (2002) argues that sentences such as 'The road runs along the coast' are manifestations of static and atemporal representations, and as such, they *contrast* with sentences such as 'The athlete runs along the coast', whose semantic profile includes actual motion along a path. It appears that theories of comprehension advocating dynamic representations (including simulation) may be better suited to account for the way people comprehend fictive motion, and the way this has been shown to affect reasoning about time (see also Matlock 2004).

## 4.3 Conclusions

The results of all our experiments support the general idea that abstract domains—those many things that we as human beings seem to grasp without being able to touch—are understood through analogical extensions from richer, more experience-based domains (Boroditsky & Ramscar 2002; Boroditsky 2000; Clark 1973; Gibbs 1994; Lakoff & Johnson 1980a). In particular, we have shown that people's thinking about the 'passage' of time is closely linked to their thinking about the way real objects move in space. It appears that when people engage in particular types of spatial-motion thinking (be it thinking about train journeys or horse races), they may also be unwittingly and dramatically affecting the structure of the representations they use to think about time. Further, and contrary to the very strong embodied view, our results suggest that abstract thinking is built on our representations of experience-based domains, and that these representations are functionally separable from those directly involved in sensorimotor experience itself.

Our results also suggest that representations of both time and fictive motion share a common base and ancestor: actual motion. Moreover, because static spatial ideas and temporal understanding have no link to one another other than through their common ancestor, it seems reasonable to assume that thinking about one or another abstract 'child' domains involves some activation of the 'parent', or of some more general abstract idea of motion extracted from and shared with the parent. This seems the most parsimonious explanation for why comprehending a fictive motion sentence in the absence of real motion can subtly influence people's understanding of time: Comprehending a fictive motion sentence appears to recruit the same dynamic representations that are used in conceptualizing actual motion, and these in turn affect the representations underpinning our ideas about time. The idea that real motion is involved seems further underlined by the last experiment described, which showed not only that fictive motion affects temporal understanding, but also that the 'direction' of fictive motion could be manipulated to create a corresponding effect on the 'direction' of temporal understanding.

Metaphor and analogy allow people to go beyond what can be observed in experience, and to talk about things they can neither see nor touch. They allow us to construct an understanding of a more abstract world of ideas. The results we describe here add credence to the widely held belief that abstract ideas make use of the structures involved in more concrete domains. Moreover, insofar as these results suggest that it is our ways of *talking* about concrete domains that seems to be at the heart of this process, they lend support to the notion that abstract ideas can be constructed and shaped not just by language, but by particular languages (Boroditsky 2001). Further, these results suggest that the human conception will not easily be partitioned into neat compartmentalized domains. Abstract ideas may take their structure from more experiential domains, but insofar as they retain the links with their siblings, these data suggest they also retain links to their parents. It remains an open and intriguing question whether, and to what extent, our knowledge of the abstract world can feed back and shape our understanding of matters that appear, on the surface at least, to be resolutely concrete.

## Acknowledgements

The authors would like to thank Amy Jean Reid, Michael Frank, Webb Phillips, Justin Weinstein, and Davie Yoon for their heroic feats of data collection.

# Section II
# From Embodiment to Abstract Thought

As the notion of embodiment enters the mainstream, it is tempting to assume we all agree—not only that embodiment exists, but also that it leads to certain mental processes and representational states. The four chapters in this section demonstrate that this is not the case. Even among those who admit a role for embodiment, there are fundamental disagreements about what embodiment is and what it implies about the nature of human thought.

This debate is often framed in terms of extremes. At one end is completely abstract thought, as exemplified by the Universal Turing Machine—an idealized mathematical processor that manipulates purely formal symbols. At the other end is absolute embodiment. The protagonist in Jorge Luis Borge's short story 'Funes el Memoriso' provides an example of the perfectly embodied thinker. As a result of an accident, he lost the ability to generalize and so was

almost incapable of general platonic ideas. It was not only difficult for him to understand that the generic term dog embraced so many unlike specimens of differing sizes and different forms; he was disturbed by the fact that a dog at three-fourteen (seen in profile) should have the same name as the dog at three-fifteen (seen from the front).

Where along this continuum, then, does human thought actually lie?

Unlike the Universal Turing Machine, we have physical bodies through which we perceive and act in a spatially extended world. Common definitions of embodiment (e.g. 'connected to the sensorimotor world') seem to admit even this obvious sense of embodiment. That cognition and language are embodied in this sense is nearly undeniable (although some philosophical idealists, such as Berkeley, have in fact denied it). Of course our concepts are learned through, our memories formed from, and our immediate thoughts influenced by direct perceptual experience. We are clearly unlike the completely disembodied Turing Machine. This does not by itself, however, entail that we are, like Funes, incapable of abstract thought.

A slightly stronger claim is that parts of our perceptual and motor systems are embodied in that their functions are to track spatial relations. Pylyshyn's (1989; 2001) Fingers of Instantiation are a good example—their function is to simultaneously track objects as they move through space. Likewise Ballard's deictic pointers (Ballard, Hayhoe, Pook, & Rao 1997) function so as to engage and disengage attention toward particular spatial locations. Given that we have material bodies through which we perceive and act in a spatially extended physical world, it is not surprising that we should have perceptual and motor mechanisms tailored for the task.

An even stronger claim is that higher cognition is shaped by experience in a spatial environment. The idea is that living in a spatial world influences, or even determines, how we think and talk. This influence may be relatively passive. For example, Carlson describes the way spatial words like 'front' are understood in terms of one's direction of movement or the way we typically manipulate objects (by their handle, lid, etc.). Or it can be active. For example, perceptual and motor processes can be recruited to offload part of the cognitive work required for a given task. Landau et al. develop such an argument. Specifically, they test whether deictic pointers are used to offload the work of perceiving and remembering spatial locations in a block construction task. This sense of embodiment—as a way to shoulder cognitive load—is relatively uncontroversial, although the amount of cognition it explains remains contentious.

A more radical view is that memory itself—not just attention, perception, or movement—is built upon the body and the world. This exceeds the claims, discussed in the previous paragraph, that the body or world can substitute for memory. Instead, it is the idea that even internally stored memories consist of physical components. Huttenlocher, Lourenco, and Vasilyeva adopt a weak version of this view. Specifically, they argue that children's memories for spatial locations include the gross topological relation of the child to the space. If spatial memories were encoded in the most abstract, observer-independent fashion, they would include only the spatial relations of objects to each other. This apparently is not the case.

Others have gone so far as to claim that the body and world are not only incidentally encoded in memories for space, but are essential elements of representations for even the most abstract notions. Barsalou's Perceptual Symbol Systems framework (1999), discussed both by Carlson and by Lipinski, Spencer, and Samuelson, is an example of what might be called 'strong embodiment.'

No one in this volume takes up the torch for what Landau, O'Hearn, and Hoffman (following Clark 1999) call 'radical embodiment'. In its strongest form, this is the view that dynamical, embodied mechanisms are all that is needed to explain intelligent behavior. Thus, internal representations of any sort are superfluous. All four chapters countenance internal representations. Although there are significant disputes over how abstract these representations are, the various ways they are embodied, and their other characteristics (are they static or dynamic? discrete or continuous? arbitrary or iconic?), nobody here categorically denies the existence of internal representations.

Both Carlson and Lipinski et al. address the relation between spatial language and spatial concepts, but they take different approaches. Carlson asks what relation holds between spatial language and the properties of real space. She points out that most research has considered how language maps onto space. For example, to say one object is in front of another requires reference to certain aspects of space (origin, direction) but not others (distance). Yet, when people interpret these terms, they actually take these other aspects of space into account. For example, distance may become important when the goal is to pick an object up. This suggests that spatial language may derive meaning from a richer web of perceptual information than is logically necessary. It also lends credence to the notion that abstract concepts are understood by simulating or re-experiencing physical interactions (*à la* Barsalou).

Lipinski et al. also are interested in the extent to which spatial language and spatial concepts overlap, but they ask about the relation between spatial language and remembered space. They demonstrate that people exhibit similar biases and response variability in both linguistic and non-linguistic tasks involving remembered locations. However, these two aspects of spatial processing (i.e. linguistic and non-linguistic) may not be so easily separated. Spencer et al. consider the non-linguistic task to be a test of spatial working memory because responses vary as a function of delay. They consider the linguistic task to be a test of spatial language because it requires a verbal response. Yet the linguistic task has exactly the same time-dependent structure as the nonlinguistic task. Therefore, both seem to involve spatial memory. Still, there is a variety of converging evidence to suggest that spatial language and non-linguistic spatial cognition use the same reference frames (e.g. Majid, Bowerman, Kita, Haun, & Levinson 2004), so their claim may well be correct.

Spatial frames of reference are hypothetical constructs that explain variation in performance on linguistic and non-linguistic tasks—explaining, for example, why some people give directions in terms of 'left/right' and not 'east/west'. One way to classify these frames is according to their origin or where they are centered. They are commonly divided into egocentric and allocentric frames. The coordinates of allocentric frames can be either object-centered or absolute. Frames of reference can be classified according to their units or how they are encoded. We generally assume that egocentric, object-centered, and absolute frames of reference are all encoded in terms of direction and relative distance—roughly, vectors in Euclidean space. However, the frames of reference Piaget ascribed to infants were coded in terms of the child's reach, not in terms of a motion-independent vector. These could be considered 'kinesthetic' reference frames, and it is possible that adults use them in some situations, such as finding the gearshift while driving. We also generally assume that only egocentric frames of reference are viewer-dependent.

Several chapters examine the properties of spatial frames of reference in some detail. Huttenlocher et al. argue that, in some circumstances, even the object-centered or relative frames of reference used to locate objects are view-dependent—they incorporate the location of the observer. Carlson suggests that that the frames of reference used in spatial language are parameterized according to both the task goals and object characteristics. In particular, whether and how distance is taken into account depends on one's goals and on the functions that an object can serve. The evidence presented by these authors challenges an implicit assumption about allocentric frames of reference—namely, that they are objective and observer-independent. On the contrary, they appear to be somewhat subjective, observer-dependent, and goal-related.

This begs the question of whether these reference frames have a psychologically reality for the average person, or are merely notational shorthand that is useful for psychologists. The chapters in this section treat frames of reference as properties of the subject, not just theoretical constructs. This naturally presupposes a certain level of spatial representation, for what else would psychologically real spatial frames be if not ways to represent spatial relations? Proponents of radical embodiment may prefer to think of them as theoretical shorthand, but this would require a new theory of spatial cognition and language that eschews spatial frames altogether. This would seem to be a formidable challenge, and might explain why none of the authors has stepped up to defend radical embodiment in its strongest sense.

# 5

# Perspectives on Spatial Development

JANELLEN HUTTENLOCHER, STELLA F. LOURENCO,
AND MARINA VASILYEVA

The ability to encode the locations of objects and places, and to maintain that information after movement, is essential for humans and other mobile creatures. While it has long been recognized that understanding spatial coding and its development is important, there are still problematic issues that require conceptual clarification and further research. In the longstanding view of Piaget (e.g. Piaget & Inhelder 1967[1948]), spatial coding in early childhood is quite impoverished. He believed that space is not initially conceptualized independently of the observer. The idea was that distance and length are coded in terms of reach rather than as features of stimuli themselves, and that location information is only preserved from an initial viewing position. In this case, there must be profound developmental changes, since adults clearly represent length and distance, and can determine how movement affects their relation to spaces that are independent of themselves. Findings in the last decade, some of them from our lab, provide strong reasons to change earlier views. In this chapter, we consider recent findings and their implications for the understanding of spatial development.

## 5.1 Coding distance in relation to space

The limits of the Piagetian view can be seen in a series of experiments by Huttenlocher, Newcombe, & Sandberg (1994). They found that, even at an early age, distance is coded independently of the observer. They had toddlers find an object after watching it being hidden in a narrow 5-foot-long sandbox. The child stood on one side of the box and the experimenter stood opposite, hid a toy, and then moved away from the box. Prior to indicating the location of the toy, the child was turned around by the parent to break gaze. Children as young as 16 months were quite accurate, showing that they used distance in coding object location. This finding suggested that distance might be coded relative to the environment rather than to the child him- or herself. Further support for this view was obtained when it was found that toddlers could locate the hidden object even if they were moved to one end of the box after watching the hiding and before the retrieval.

Newcombe, Huttenlocher, Drummey, & Wiley (1998) further explored this issue by having children move around to the other side of the box after the hiding and before the retrieval event. Children were still able to indicate where the hidden object was located, and they showed systematic bias towards the center of the space. That is, they located objects as slightly, but significantly, nearer to the center than the true location. This pattern of responding showed not only that young children code location relative to the box, but also that they seem to be sensitive to the geometric properties of a space, in particular to the center of the box. Together, the findings clearly indicate that young children code location relative to the outside environment, not simply relative to themselves.

## 5.2  How information is maintained after movement

Given that even toddlers code object location relative to outside environments (enclosed spaces), questions arise as to how they maintain this information as they move to new positions. In many tasks involving movement, the location of an object can be coded egocentrically, relative to the child's initial position. During the process of moving to a new position, the viewer's location relative to an object can be continuously updated. Alternatively, viewers might not track their movement, but rather might code the object's relation to the space without reference to themselves. The first of these possibilities (tracking changes during movement) is based on initial coding in relation to the self, and, in this sense, presents less of a departure from Piaget's views of early spatial representations. However, recent findings indicate that toddlers maintain information about object location even when they cannot track their own movements relative to a target.

A series of studies found that toddlers can retain object location when a disorientation procedure is used. Hermer & Spelke (1994; 1996) adapted the disorientation procedure from earlier work with rats. Cheng (1986) placed rats inside a rectangular box and showed them that food was hidden in a particular location. After observing the hiding of the food, the rats were placed in another dark box and moved around (i.e. disoriented) prior to being allowed to search for the hidden food. In the parallel task given to humans, young children were placed in a small rectangular room and shown an object being hidden in a corner. They were then disoriented by being turned around several times with their eyes covered prior to searching for the hidden object. This procedure ensured that children could not simply maintain the location of the object by tracking their movement in relation to the hiding corner.

In a disorientation task, locating a target object involves establishing its position relative to the space itself. The spatial characteristics used might include landmarks, geometric cues, or both. The geometric cues of a rectangular room include corners that can be distinguished from each other on the basis of the relative lengths of the walls—one pair has the longer wall to the right of the shorter wall, whereas the other pair has the longer wall to the left of the shorter wall. Studies involving disorientation provide striking evidence that toddlers use the geometric cues of an enclosed space in searching for a hidden object. Both Cheng (1986) and Hermer & Spelke (1994; 1996) found that rats and toddlers searched in a geometrically appropriate location after disorientation—either the correct corner or the equivalent corner diagonally opposite it. The results obtained by Hermer & Spelke are shown in Figure 5.1. Interestingly, geometric cues were used to the exclusion of other information. In particular, landmark information (e.g. the color of a wall), which could potentially distinguish the correct corner from the geometrically identical corner (e.g. longer *blue* wall to the left of shorter white wall vs. longer *white* wall to the left of shorter white wall), was ignored, and search was based solely on geometry. These findings led the investigators to posit that geometric sensitivity was a modular ability.

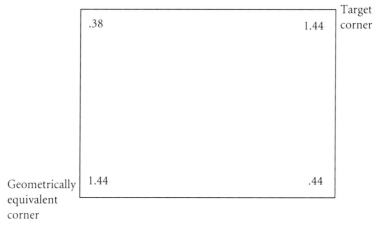

FIGURE 5.1. Average number of responses at each corner in the Hermer & Spelke (1996) study

## 5.3 Modularity

The proposal by Spelke and colleagues (e.g. Hermer & Spelke 1994; 1996; Wang & Spelke 2002) was that early geometric sensitivity involves a specific ability to code the geometry of spaces that surround the viewer. Several issues must be addressed to evaluate this claim. It has been argued that the critical information about the location of a hidden object involves the child's initial direction of heading. Finding the hidden object then would require recovering the initial heading, reorienting by 'aligning the currently perceived environment with a representation of its geometric and sense properties' (Hermer & Spelke 1996: 208). This claim has been tested in our recent research described below. Further, it has been argued that children do not conjoin geometric cues and landmark information to determine object location. That is, Hermer & Spelke posited that geometric processing does not admit the use of non-geometric information such as landmarks. There is accumulating evidence, however, that both animals and human children *do* combine landmark and geometric information on disorientation tasks under certain conditions. Learmonth, Newcombe, & Huttenlocher (2001) showed that varying room size affects whether landmarks are used. Indeed, children who used landmarks to disambiguate geometrically equivalent corners in a large room ignored landmarks in a small room (Learmonth, Nadel, & Newcombe 2002).

Studies with different species of animals also show the use of landmarks to locate an object. As is the case with toddlers, the use of landmarks depends on the particular context. Rhesus monkeys, for example, incorporate geometric and landmark information only when large featural cues are used (Gouteux, Thinus-Blanc, & Vauclair 2001). In other cases, animals show a robust sensitivity to landmarks in disorientation tasks. For example, Vallortigara, Zanforlin, and Pasti (1990) found that chicks not only used landmarks to distinguish between the target and the geometrically equivalent corners, but actually preferred landmarks when these were in conflict with geometric cues. In reviewing existing work on human and non-human animals, Cheng & Newcombe (2005) concluded that, while geometric cues are prepotent in most cases, landmarks can be incorporated into the coding of spaces.

## 5.4 The breadth of geometric ability

While there have been extensive discussions of modularity (i.e. of ignoring non-geometric information), other important questions have not received much attention. Notably, the generality of children's sensitivity to geometric information has not been fully explored. The studies by Spelke and colleagues used rectangular spaces and tested children inside those spaces. However, to characterize the nature of early geometric sensitivity, it is important to investigate if toddlers can code spaces of different shapes and if they can find an object from other viewing positions, namely, from outside as well as from inside a space.

Initially, evidence was presented to support the view that early geometric sensitivity was a narrow ability specialized to deal with surrounding spaces (Wang & Spelke 2002). In support of this, Gouteux, Vauclair, & Thinus-Blanc (2001) found that 3-year-olds were unable to use geometric information to locate an object in the corner of a rectangular box when they were outside that space. The task used by Gouteux et al., however, involved movement of the space relative to a stationary child rather than movement of the child relative to a stationary space. Yet it is well known that movement of a spatial layout is not equivalent in difficulty to movement of the viewer (Huttenlocher & Presson 1979; Simons & Wang 1998; Wraga, Creem, & Proffitt 2000), and recently Lourenco & Huttenlocher (2006) showed that young children's search accuracy varied as a function of the type of disorientation procedure (i.e. viewer versus space movement). Further, rather than using a constant location for the hidden object over trials, as in previous studies, Gouteux et al. varied the hiding location across trials, which may have resulted in perseverative errors.

In a series of experiments, Huttenlocher & Vasilyeva (2003) examined the extent to which children's coding generalized to different-shaped spaces and to different viewing positions (i.e. inside versus outside). The task was one where children were moved relative to a stationary space and the location of the hidden object was held constant. Children were tested in a room the shape of an isosceles triangle (as shown in Figure 5.2). One of the corners was unique in angle, with equally long walls on each side. The other two corners were equal in angle, with walls that differed in length; one of the corners had the long wall on the right and the short wall on the left, and the other had the long wall on the left and the short wall on the right, as in a rectangular space. The procedure was parallel to that followed in previous studies with a rectangular room.

The results showed that performance in the triangular room was comparable to that in a rectangular room. That is, the overall success rate was 70%, well above the chance level of 33% for a triangular space. Hermer & Spelke (1994; 1996) had found that the success rate in a rectangular room was 78%, where chance was 50%. Our results, like those of Hermer & Spelke, indicate that children had maintained information about the hiding corner even after disorientation. In a rectangle, the four angles are equal, and the cues that distinguish the corners consist of differences in the lengths of walls that form it. In the isosceles triangle used in our study, there is an additional cue—one of

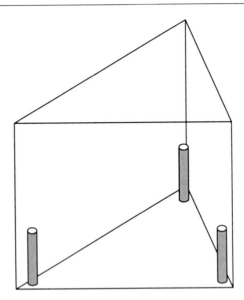

FIGURE 5.2. Triangular room used in the Huttenlocher & Vasilyeva (2003) study

the corners is unique in angular size. If children had used angular information in addition to side length, accuracy at the unique corner might have been greatest. Since performance was equivalent for all of the corners, it may be that children used information either about the equal length of the sides or about the angular size, but not both. Further, when the object was hidden in one of the two equal-sized corners, children might have been more likely to confuse these corners with one another than with the unique corner. However, we found no evidence of a difference in task difficulty depending on the corner of hiding (see also Lourenco & Huttenlocher 2006). This suggests that children rely on information about the relative lengths of adjacent walls in representing both triangular and rectangular spaces (see also Hupach & Nadel 2005; for review, Lourenco & Huttenlocher 2007).

We also tested children who were positioned outside of triangular and rectangular spaces. Because these experimental spaces were only 6 inches deep, children could see and reach into them from outside. The shapes were surrounded by a large round fabric enclosure high enough to prevent the use of other cues such as those from the experimental room. The procedure in these experiments was similar to that in experiments where the child was inside, except that the disorientation procedure involved the parent holding the child (whose eyes were covered) and walking around the space.

Note that when a space is viewed from outside, the lengths of the sides relative to the child and appearance of corners depends on where along the perimeter of the space the child is located. For example, a particular corner of a triangle, viewed from outside, may have the long side to the left and the short side to the right, joined by an angle of 70°. From the opposite vantage point, however, the same corner has the short side to the left and the long side to the right, joined by an angle of 290°. See Figure 5.3 for an illustration. Hence, if children rely on a particular view of the original hiding location, the existence of multiple perspectives might greatly increase task difficulty, since the look of the hiding corner from the initial perspective may be very different from its appearance after disorientation.

When children were positioned outside a triangular space, they were correct on 56% of trials. While this performance level is significantly above chance (33%), it is lower than when children were tested inside the space (70%). When children were positioned outside a rectangular space, they searched in one of the two geometrically correct corners on 69% of trials. Again, this performance was well above the 50% chance level, but success was somewhat lower than in the original Hermer & Spelke study, where toddlers were correct on 78% of trials. Thus, for both the triangular and rectangular spaces, the task appears to be more difficult when children are tested from outside.

Based on our results, it is clear that toddlers' geometric ability is more general than described in the original work on disorientation. Toddlers can locate a hidden object after disorientation in a triangular as well as in a rectangular space. Further, toddlers are not restricted to spaces that surround them; they can also

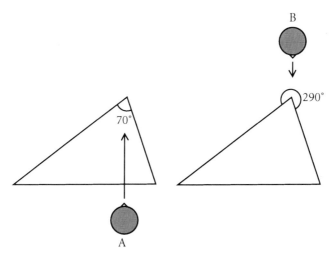

FIGURE 5.3. Alternative views (A and B) of a corner

code object location when they are outside a space. The fact that the task differs in difficulty depending on the position of the child (outside versus inside) indicates that the viewer is involved in the representation of the space. If the coding had been independent of the viewer, the task would have been equally difficult regardless of the child's position. We return to this issue later in the chapter.

## 5.5  Representation of the space

Major questions remain as to what the disorientation studies reveal about the way space is represented and how hidden objects are retrieved. As we have noted, when viewers who are not disoriented change position relative to a space, they can track their changing relation to the particular portion of the space where the object is hidden. That is, they can code the amount and direction of a change in position; for example, if a person turns 180°, the object that was in front will now be behind him or her. The disorientation procedure, however, prevents such tracking, breaking the link between the viewer's original position in a space and the location of a target object.

There is more than one possible way for a person to code the location of an object in relation to a space so as to be able to succeed on disorientation tasks. One way to code location is to represent the portion of the space that includes the target object as seen from the original viewpoint (e.g. the corner with the longer wall to the left and the shorter wall to the right). In a sense, the strategy of coding location from one's initial position is similar to egocentric coding of space such as Piaget proposed. However, unlike Piaget's proposal, this representation would have to be maintained when the viewer moves to new positions. Finding the object after disorientation in this case would involve searching for a view that matches the starting viewpoint or 'initial heading', as Hermer & Spelke proposed.

There is another possible way to code location. It would be to code the shape of the whole space with the hidden object in it. The space might be represented in terms of the internal relations among its parts. This representation would be independent of the viewer's original heading towards a particular portion of a space. In such a conceptualization, viewer perspective might be relative to the entire space (as inside or as outside), or might not involve a particular viewing position at all. In either case, no matter what portion of the space the viewer faces after disorientation, the relation between the viewer's position and the location of the hidden object would be known without searching for the original heading.

While previous studies involving disorientation have focused on performance accuracy, Huttenlocher & Vasilyeva (2003) noted that further insight could be gained by examining how children search for a hidden object. They studied the problem by exploring the behaviors children engage in when finding a hidden object following disorientation. Specifically, children's search behavior was examined to determine if they surveyed the various corners of the space where the object might be hidden, or if they went directly to a particular corner.

If children rely on coding their original view of the portion of the space that contains the hidden object, they would have to survey the space by moving or looking around to find that original view after disorientation. That is, they would have to examine various potential hiding locations to find the one that matches their original view of the space. If, on the other hand, children represented the entire space in terms of the internal relations among its parts, then no matter what their position was after disorientation, they would know the relation between their current position and the hiding location. Thus, they would be able to find the hidden object without having to recover the particular perspective that matches their original view.

To examine children's search behaviors, a video camera that recorded the course of the experiment was mounted to the ceiling of the room. These pictures made it possible to determine whether or not children had surveyed the space following disorientation. If children turned their head, torso, or whole body prior to searching for the hidden object, they were classified as having surveyed the space. When children were tested inside the triangular space, it was usually, but not always, possible to make a classification. On 23% of trials, it was not possible to determine the type of behavior because children were somehow occluded in the video or their movements were too subtle. However, on 69% of trials, children clearly went directly to a particular corner without surveying the space. This finding is not likely to reflect a failure of the disorientation procedure, since work with rectangular spaces with geometrically equivalent corners shows that children do not distinguish between equivalent corners when disoriented. On only 8% of the trials did children actually look around at the different corners. Their success on a given trial was statistically equivalent whether or not they had attempted to survey the space.

In the case when children viewed a space from outside, it was easier for the investigator to determine if the children were surveying the alternative corners than in the case when they were inside. Indeed, it was possible to classify children's search behavior on all trials. For the triangular space, they went directly to one of the corners without surveying the space on 89% of trials. On the other 11%, they looked around at more than one corner. Again, the children did not perform better in those rare cases when they actually surveyed the space. Results were parallel with rectangular spaces: on 86% of the trials they did not survey the space, and on 14% of the trials they did survey the space. The answer to the question of whether children have to search for their original heading is clear: they succeed on the task without doing so. Hence, we concluded that children code the entire space, not just the hiding corner as it appeared from the original viewpoint.

## 5.6  Representation of the viewer

Recall that Huttenlocher & Vasilyeva (2003) found different success levels depending on whether children were inside or outside a space. This finding suggests that viewer perspective is incorporated in the spatial representation. In fact, if viewer position were not coded (i.e. if the coding were strictly space centered), then the position of the viewer could not affect task difficulty. Yet the difficulty of inside versus outside tasks could differ for other reasons as well. That is, differences in task difficulty do not necessarily imply that the viewer is included in the spatial representation. Let us consider some alternative explanations.

One possibility, described above, is that when an enclosed space is viewed from outside, the look of a given corner differs depending on where the viewer stands along the perimeter of the space. Thus, the greater difficulty of outside tasks might reflect the fact that there are multiple views on the corners. A second possibility is that the conceptualization of a space is more abstract when it is coded from inside than when it is coded from outside. That is, from inside, the entire space is not seen all at once and hence must be 'constructed' by the viewer; from outside, the entire space can be seen at once and hence the coding might be more 'iconic'. The constructed space, coded from inside, might be more consistent with maintenance of information about the hiding location after disorientation. A third possible reason for the observed difference in difficulty concerns the particular experimental conditions of the study. That is, the size of the spaces differed for the inside and outside tasks; the space used in the inside condition was considerably larger than the space used in the outside condition, and this could have led to differences in task difficulty.

Earlier findings by Huttenlocher & Presson (1973; 1979) are relevant to evaluating these alternatives. In that work, children were presented with object location tasks both inside and outside the same enclosed structure. The structure was 6 feet high, so there was no problem of multiple perspectives in the outside condition. Furthermore, the space could not be seen at once, either from inside or outside. Finally, the very same space was used in both tasks. Nevertheless, there was a difference in task difficulty parallel to the one reported by Huttenlocher & Vasilyeva (2003): the outside task was harder. Thus, the Huttenlocher & Presson work suggests that differences in performance on inside versus outside tasks are not likely due to any of the alternatives discussed above.

Having tentatively set aside these alternative hypotheses, let us consider a possible explanation for why the outside version of the task is more difficult than the inside version. We hypothesize that this difference may reflect variation in the distinctiveness of the critical information about the enclosed space when the viewer is inside versus outside (see Figure 5.4). When facing the space from outside, the whole space is in front of the viewer and all the potential hiding corners are in a frontal plane; since these positions are similar, they are potentially confusable. In contrast, when a viewer is inside, the potential hiding corners are not all in the frontal plane: one is in front, one behind, one to the left, and one to the right. These positions are more differentiated relative to the viewer, perhaps making the inside task easier.

The findings thus far indicate that toddlers represent simple enclosed shapes in terms of the internal relations among all of the parts (i.e. the whole space). The findings also show that viewer perspective is incorporated into the representation, but not in the way most commonly discussed in the spatial cognition literature. In its common use, the notion of viewer perspective refers to the relation of an individual to a particular portion of space that he or she faces. That is, such coding is taken to involve what the viewer sees from a fixed position (e.g. Piaget & Inhelder 1967). Another sense of viewer perspective would involve the relation of an individual to an entire space, namely, as being inside or outside that space. It would appear that it is in this sense that the viewer is incorporated into the representation of the space, according to the initial findings of Huttenlocher & Vasilyeva (2003).

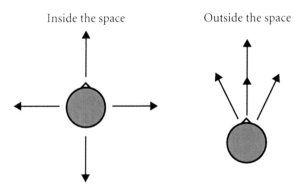

Inside the space          Outside the space

FIGURE 5.4. Schematic representation of inside and outside perspectives

## 5.7 Is the viewer omni-present?

Having obtained evidence that viewers code their relation to the entire space, the question arises of whether the viewer is always included in the spatial representation or whether there are conditions where the viewer is not represented. It would seem that if viewer location in a space were varied even more widely than in the Huttenlocher & Vasilyeva study, its position relative to the hidden object might become more difficult to determine, and toddlers might instead represent object location purely in terms of position relative to the space. In previous disorientation studies, the viewer has remained either inside or outside the space for all phases of the task (hiding, disorientation, and retrieval) so that the viewer's position relative to the entire space was constant.

Recently, Lourenco, Huttenlocher, & Vasilyeva (2005) conducted a study involving both disorientation and translational movement. After the hiding and before retrieval, the child was not only disoriented but also moved relative to the entire space, that is, 'translated' from inside to outside or vice versa. Toddlers either stood inside the space during hiding and outside during retrieval, or outside during hiding and inside during retrieval. The sequence of disorientation and translation was also varied: disorientation occurred either before or after the child was moved into or out of the rectangular space. Thus four groups were studied. For two groups, the children were moved from inside to outside; for the other two, the children were moved from outside to inside. Each of these groups was subdivided according to whether translation or disorientation occurred first. We also conducted a control experiment involving a disorientation task like that in our previous studies where children remained either inside or outside the space throughout the entire procedure. In all conditions, the space was rectangular, with identical containers in each corner serving as potential hiding locations. This space was large enough that a child and adult could stand inside it comfortably and small enough that the entire space could be seen from outside.

Since the conditions were all identical except for the movements of the participants, if children coded the location of the hidden object solely in terms of its position relative to the enclosed space, performance would be the same for all conditions. If the viewer is involved in the representation, then changes in viewer position might affect task difficulty. In fact, if viewers were unable to take account of their changing relation to the space, they would not find the object. Let us consider further what is involved in coding viewer perspective in tasks that include translational movement (both into and out of the space) and disorientation.

A viewer's changing relation to a hidden object cannot be tracked when that viewer undergoes disorientation. However, the relation to an object can be tracked when a viewer undergoes translational movement. The difficulty of a task that involves both disorientation and translation might depend on the order of these two transformations. In particular, when translation precedes disorientation, toddlers can track and update their relation to the target corner as they move from inside to outside or vice versa. Disorientation after translation then might be no more difficult than if a previous translation had not occurred. However, when disorientation precedes translation, the situation is quite different. As we have shown above, when toddlers are disoriented, they rely on a coding of the entire space. Therefore, when translation occurs after disorientation, toddlers would need to transform the entire space, not just a particular portion of the space. Transforming information about the entire space is more difficult than transforming information about a single object, possibly making the task where disorientation precedes translation quite difficult.

The results indeed showed a very strong effect of the order of translation and disorientation. When translation preceded disorientation, toddlers performed as well as in other work with disorientation alone. When they searched for the hidden object from inside, having been moved into the space, they chose one of the two geometrically appropriate corners 75% of the time. When they searched for the object from outside, having been moved out of the space, they chose an appropriate corner 64% of the time. As we hypothesized, performance on this task was similar to that in our control conditions with no translation (80% correct from inside and 66% correct from outside). In contrast, when disorientation preceded translation, performance was at chance, both when the child searched from outside after the hiding and disorientation had occurred from inside and vice versa. The results obtained by Lourenco et al. (2005) indicate that the viewer has difficulty ignoring his or her own position relative to the space even when it prevents successful performance.

Let us consider why the viewer's position is critical. The reason, we believe, is that since the task involves retrieving the hidden object, a link between viewer and object is required. Even though disorientation tasks disrupt the link between the viewer and the object, children can succeed on some of these tasks, namely, tasks where the viewer and object bear a common relation to the space. As noted above, when translation precedes disorientation, viewers can track their movement into or out of the space relative to the hidden object during the translation process. Since their relation to the entire space remains constant in the process of disorientation, they maintain a common relation to the entire space such that they can infer the link between the viewer and the object, dealing with disorientation as if translation had not occurred.

However, when disorientation occurs first, they code the relation to the entire space. Hence, during translation, viewers must track the change in their relation to the entire space in order to establish their link to the hidden object. If viewers were able to transform their relation to the entire space with the object in it, they could infer their relation to the object itself. Toddlers possibly fail this task because it is difficult for them to transform the whole spatial array (see also Lourenco & Huttenlocher 2007).

## 5.8  Summary and conclusions

In this chapter, we have described recent advances in our understanding of spatial development. One advance has been to show that very early in life children possess a sensitivity to the geometry of their spatial environment. This ability is not easily demonstrated in natural contexts because it is difficult to determine whether spatial features actually have been coded. That is, viewers often can track their changing relation to a particular object or place as they move, so they need not represent the spatial environment itself. Recently, however, methods have been developed to prevent viewers from tracking changes in their relation to a target location, making it possible to determine if geometric cues are indeed coded. In particular, a disorientation technique has been introduced in which individuals are moved in space in a way that disrupts their ability to track the hiding location. Thus, this technique makes it possible to investigate whether viewers represent the relation of a hidden object to particular features of an enclosed space.

Using the disorientation procedure, Cheng (1986) showed a sensitivity to the geometry of a simple enclosed space in rats, and Hermer & Spelke (1994; 1996) showed a parallel sensitivity in very young humans. That is, toddlers code geometric information about the location of a hidden object relative to a space, allowing them to find the object after disorientation. However, the nature of the underlying representation was not initially clear. The investigators posited the existence of a cognitive module, in which only geometric properties of enclosed spaces are processed. It was posited that this module allows a viewer to locate a hidden object by re-establishing his or her initial heading towards the hiding corner. However, Huttenlocher & Vasilyeva (2003) found that when toddlers retrieved an object after disorientation, they did not engage in a search to re-establish their initial heading. Rather, they went directly to a particular corner, suggesting that they had coded the entire space, not just their original heading.

Another recent advance in our understanding of spatial development has involved obtaining evidence that incorporating the viewer into spatial representations may be obligatory, at least when action in relation to a space is involved. The term 'viewer perspective' here differs from the traditional notion that it involves an individual's position relative to a particular portion of a space (i.e. target location). The sense of perspective invoked here involves the coding of viewer position relative to an entire space. The evidence of this form of viewer perspective is that task difficulty is affected by viewer position inside or outside the space. Lourenco et al. (2005) have presented evidence that both forms of perspective may be essential elements of spatial representation. In that study, viewer position was varied in relation to a portion of the space as well as to the entire space by both disorientating and translating the viewer. If viewers did not code their perspective, neither of these operations, nor the order of their presentation, should have affected task difficulty. In reality, the order of disorientation and translation had a large effect on task difficulty. If children were translated (from inside to outside the space or vice versa) before disorientation, the task was as easy as if they had not been translated at all. However, if translation occurred after disorientation, the children performed at chance. The fact that viewer movement had such a significant influence on task difficulty suggests that viewers attempt to code their own perspective in relation to enclosed spaces.

In short, two types of perspective coding have been identified. These forms of coding may coexist, but the one that underlies particular behaviors may depend on the task. For example, when viewers remain stationary, only perspective on a portion of space may be relevant. When viewers move in a manner that can be tracked, they can still use their perspective on a portion of the space, updating this perspective as their relation to the target changes. However, if viewers are disoriented and cannot track their relation to the target, they may code their perspective relative to the entire enclosed space. When the viewer and object are both coded with respect to a commonly defined space, even young children can infer their own relation to the object. On some tasks, however, this inference requires transforming the viewer's relation to the entire space, which may be very difficult for young children.

# 6

# It's in the Eye of the Beholder: Spatial Language and Spatial Memory Use the Same Perceptual Reference Frames

JOHN LIPINSKI, JOHN P. SPENCER,
AND LARISSA K. SAMUELSON

Representations of words are often viewed as discrete and static, while those of sensorimotor systems are seen as continuous and dynamic, a distinction mirroring the larger contrast between amodal and perceptual symbol systems. Spatial language provides an effective domain in which to examine the connection between non-linguistic and linguistic systems because it is an unambiguous case of linguistic and sensorimotor systems coming together. To this end, we reconsider foundational work in spatial language by Hayward & Tarr (1995) and Crawford and colleagues (2000) which emphasizes representation in the abstract. In particular, we use a process-based theory of spatial working memory—the Dynamic Field Theory—to generate and test novel predictions regarding the time-dependent link between spatial memory and spatial language. Our analysis and empirical findings suggest that focusing on the processes underlying spatial language, rather than representations *per se*, can produce more constrained theories of the connection between sensorimotor and linguistic systems.

## 6.1 Introduction

A fundamental issue in the study of language is the relationship between the representations of words and sensorimotor systems that necessarily operate in the real world in real time (Barsalou 1999; Harnad 1990). Representations of words are typically viewed as discrete, arbitrary, and static, while sensorimotor systems typically trade in continuous, non-arbitrary, and dynamic representations. From a theoretical standpoint, the challenge is to understand how two such seemingly different representational formats communicate with each other (Bridgeman, Gemmer, Forsman, & Huemer 2000; Bridgeman, Peery, & Anand 1997; Jackendoff 1996). The domain of spatial language is an ideal testing ground for proposals addressing this representational gap precisely because it is an unambiguous case of linguistic and sensorimotor systems coming together.

Within the field of spatial language, the issue of representational formats has a long, rich history, from the extensive linguistic analysis by Talmy (1983), who argued that schematic representations underlie spatial term use, to more recent efforts that have examined the real-time activation of linguistic representations by sensory inputs (Spivey-Knowlton, Tanenhaus, Eberhard, & Sedivy 1998). This diversity of approaches has led to a diversity of perspectives regarding the nature of the relationship between spatial language on the one hand and spatial perception, spatial memory, and spatial action on the other hand. Some researchers contend that linguistic and non-linguistic representations overlap in fundamental ways (Avraamides 2003; Hayward & Tarr 1995; Loomis, Lippa, Klatzky, & Golledge 2002), while other researchers contend that these are distinctly different classes of representation (Crawford, Regier, & Huttenlocher 2000; Jackendoff 1996).

Although the rich literature on spatial representations has led to important insights about the nature of linguistic and non-linguistic spatial systems, the central thesis of the present chapter is that this work suffers from a heavy emphasis on static representations. This, combined with the often conceptual nature of the theories proposed in the spatial language domain, leads to theories that are under-constrained and empirical findings that can be interpreted in multiple ways. We contend that the current state of affairs warrants a new approach that emphasizes the *processes that give rise to representational states*, that is, the second-to-second processes that connect the sensorimotor to the cognitive—both linguistic and non-linguistic—in the context of a specific task. We use the term 'representational state' to contrast our emphasis on process with previous work that has emphasized static representations. A representational state by our view is a time-dependent state in which a particular pattern of neural activation that reflects, for instance, some event in the world is re-presented to the nervous system in the absence of the input that specified that event. Note that this view of re-presentation is related to recent ideas that the brain runs 'simulations' of past events during many cognitive tasks (see e.g. Damasio & Damasio 1994; for further discussion see Johnson, Spencer, & Schöner, in press; Spencer & Schöner 2003).

There are three key advantages to emphasizing the processes that give rise to representational states. First, process models are more constrained than models that focus primarily on static representations because they must specify two things: the processes that give rise to representational states as well as the nature of the representational states themselves. In our experience, handling the first issue provides strong constraints on possible answers to the second issue (Spencer & Schöner 2003). Second, theories that focus too narrowly on static representations tend to sidestep the central issue we began with: how to connect the dynamic world of the sensorimotor to the seemingly discrete world of the linguistic. By contrast, process-based theories provide useful grounding, forcing researchers to take the real-time details of the task and context seriously. Third, we contend that an emphasis on process can lead to new empirical questions and new methods to answer them. We illustrate this with a novel set of findings that probe the link between spatial language and spatial memory. These empirical efforts build upon other recent insights gained from thinking of language and cognition as 'embodied', that is, intricately connected with the sensorimotor world (see Barsalou 1999; Spivey-Knowlton et al. 1998; Stanfield & Zwaan 2001; Tanenhaus, Spivey-Knowlton, Eberhard, & Sedivy 1995; Zwaan, Madden, Yaxley, & Aveyard 2004; Zwaan, Stanfield, & Yaxley 2002).

With our broad issues now framed, here are the details of how we will proceed. First, we give a brief overview of how the link between linguistic and non-linguistic representations has been conceptualized within the domain of spatial language (section 6.2). Although these approaches are rich conceptually, they have not provided a theoretical framework constrained enough to produce critical empirical tests (section 6.3). Next, we discuss an ongoing debate about spatial preposition use that has attempted to simplify the problem of connecting sensorimotor and linguistic systems by focusing on the representations underlying spatial language (section 6.4). Although data generated in the context of this debate are compelling, the accounts that have been proposed are under-constrained. We claim that thinking about process can shed new light on such debates. Thus, in section 6.5, we apply a new theory of spatial working memory—the Dynamic Field Theory [DFT] (Spencer & Schöner 2003; Spencer, Simmering, Schutte, & Schöner 2007)—to the issue of how people activate and use spatial information in linguistic and non-linguistic tasks. We then test some novel predictions inspired by our model (section 6.6). Finally, we return to the wider literature and highlight some implications of our process-based approach as well as some of the future challenges for our viewpoint (section 6.7).

## 6.2 Two approaches to the linguistic/non-linguistic connection

A fundamental strength of language is its ability to connect abstract symbols that refer to objects in the real world to the dynamic sensorimotor systems that perceive and interact with these objects. Because spatial language brings words and physical space together so directly, it is the ideal vehicle for exploring this interaction. To date, two general approaches speak to this issue of the linguistic/non-linguistic connection in spatial language: amodal symbol systems and perceptual symbol systems.

### 6.2.1 *Amodal symbol systems*

Amodal symbol systems presume representational independence between symbolic processes like language and sensorimotor systems (Anderson 2000; Harnad 1990). The amodal view thus requires a transduction process that permits 'communication' between linguistic and non-linguistic systems. This transduction process is best described by Jackendoff's representational interface (Jackendoff 1992; 1996; 2002). Representational interfaces account for communication between different types of representation (e.g. verbal and visual) by proposing a process of schematization—the simplifying and filtering out of information within one representational format for use in another representational system (Talmy 1983). The representational interface approach ultimately permits abstract conceptual structures to encode spatial representations while still capturing the core characteristics of the symbolic view (e.g. pointers to sensory modalities, type-token distinctions, taxonomies).

There is significant empirical support for this view. Consistent with Jackendoff's representational interface, for example, Talmy (1983) showed that language uses closed-class prepositions (such as 'above', 'below', or 'near') to provide an abstracted, skeletal structure of a scene that narrows the listener's attention to a particular relationship between two objects by disregarding other available information. In the sentence 'The bike stood near the house', for example, Talmy shows that all of the specific information about the bike (e.g. size, shape, orientation) is disregarded and the bike is instead treated as a dimensionless point (Hayward & Tarr 1995). As a result of this schematization, such a linguistic representation of a relational state can be extended to a variety of visual scenes and objects without much regard to the individual object characteristics (Landau & Jackendoff 1993).

### 6.2.2 *Perceptual symbol systems*

In contrast to the transduction view of the amodal approach, Barsalou's Perceptual Symbol Systems [PSS] (1999) posits a more intricate connection between the linguistic and non-linguistic. By this view, transduction is not needed because symbols—perceptual symbols—arise from the same neural states that underlie perception. In particular, perceptual symbols arise when top-down processes partially reactivate sensorimotor areas and, over time, organize perceptual memories around a common frame. Once such a frame is established, the perceptual components of the frame can be reactivated, forming a 'simulator' that captures key elements of past experiences as well as core symbolic aspects of behavior such as productivity, type-token distinctions, and hierarchical relations. In this way, perceptual symbols are both inherently grounded in the cortical activations produced by a given sensory modality and capable of replicating the flexible, productive, and hierarchical capacities of amodal symbolic systems. Moreover, because these symbols are grounded in sensorimotor processes, they do not require pointers or transduction to become 'meaningful'.

A growing empirical literature supports Barsalou's (1999) PSS. For example, Stanfield and Zwaan (2001) argued that if symbolic, linguistic representations are integrated with perceptual systems, people should be faster to recognize visual objects described in a sentence as the similarity between the perceived object and the description increase. Consistent with this prediction, they found that people were faster to recognize an object (e.g. a vertically oriented pencil) as part of a previous sentence when that sentence matched the orientation (e.g. 'He placed the pencil in the cup') than when it conflicted (e.g. 'He placed the pencil in the drawer'). Additional evidence for the tight integration of visual and linguistic representations comes from head-mounted eye-tracking data acquired during linguistic processing tasks. Such data show that eye movements used to scan a visual scene are time-locked to verbal instructions to pick up items within that scene (Spivey-Knowlton et al. 1998). Visual information has also been shown to facilitate real-time resolution of temporarily syntactically ambiguous sentences (Tanenhaus et al. 1995)—further evidence against a hard separation between linguistic and sensory systems. Finally, work by Richardson, Spivey, Barsalou, & McRae (2003) shows that spatially grounded verbal stimuli interact with visual discrimination performance, providing additional evidence that linguistic processing can directly impact the processing of visual space.

## 6.3 Limits of the amodal and perceptual symbols system approaches

The amodal and PSS views are opposites conceptually; however, both perspectives appear to be substantially supported within the spatial language domain. This is not an acceptable state of affairs, because two opposing perspectives proposed to account for the same phenomena cannot both be correct. For instance, if the PSS view were correct, amodal symbols would be superfluous because symbolic processes would fall out of the organization of dynamic, schematic records of neural activation that arise during perception (Barsalou 1999). Thus, despite a vigorous debate and valuable empirical data on both sides, the fundamental question of how spatial linguistic and non-linguistic systems are connected remains unanswered. Further consideration suggests a critical limitation of these proposals: the amodal and PSS views rely on descriptive, conceptual accounts of the linguistic/non-linguistic connection. Though often useful at initial stages of theory development, the flexibility of conceptual accounts makes them ultimately difficult to critically test and falsify. Consequently, data collected in support of one view can, in some cases, be reinterpreted by the other view. Jackendoff (2002), for example, has explained the real-time resolution of syntactic ambiguity through visual processing (Tanenhaus et al. 1995) using characteristics of a syntax-semantics interface.

Conceptual theories are particularly problematic in the context of the linguistic/non-linguistic connection because of the complexity of the theoretical terrain: these theories must explain the process that unites spatial terms with spatial perception, memory, and action. More concretely, such theories have to specify how people perceive a scene, how they identify key spatial relations such as the relation between a target object and a reference object, how such spatial relations are remembered in the context of real-time action of both the observer and the environment, and how these relations are used in discourse to produce a verbal description sufficiently detailed to allow another person to act on that information. The conceptual theories discussed above make reference to processes involved in such situations—transduction processes on one hand, simulation processes on the other—but the formal details of these processes are lacking. Given the complexity of what these theories have to accomplish, this is not surprising.

Although a formal theory seems relatively distant at present, we can ask a simpler question: what might a formal theory of such processes look like? Barsalou's (1999) move to embrace neural reality seems particularly appealing in that it highlights possible connections among conceptual theory (e.g. the PSS view), neurally inspired formal theories (e.g. neural network approaches), and data (e.g. fMRI or single-unit recordings). Indeed, there are several neurally plausible theories of key elements of the linguistic/non-linguistic connection (e.g. Cohen, Braver, & O'Reilly 1996; Gruber & Goschke 2004; Gupta & MacWhinney 1997; Gupta, MacWhinney,

Feldman, & Sacco 2003; McClelland, McNaughton, & O'Reilly 1995; O'Reilly & Munakata 2000). Although these potential links are exciting, they are also daunting given the added complexities of dealing with a densely inter-connected and highly non-linear nervous system (e.g. Freeman 2000). For instance, how might a population of neurons that encodes a particular spatial relation link up with other populations that deal with lexical and semantic information? And how might these different populations allow their patterns of activation to mingle and integrate, while at the same time stably maintaining their own unique content in the face of neural noise and changing environments (Johnson et al. in press; Spencer & Schöner 2003; Spencer et al. 2007)?

Perhaps on account of this daunting picture, many researchers have split the linguistic/non-linguistic connection problem up into two parts: (1) what is the nature of the representations used by linguistic and non-linguistic systems, and (2) how are they connected? Within this framework, the vast majority of research has focused on the first question: the representational format used by spatial perception, action, and memory on one hand and spatial language on the other. Although, as before, this view has generated many insightful empirical findings (some of which we describe below), it has led to under-constrained theories of the representations that support performance. We contend that this is a natural by-product of emphasizing representations in the abstract, rather than the processes that give rise to representational states. Moreover, we claim that the latter approach ultimately leads to more constrained theories and, perhaps, a richer view of how the sensorimotor and the linguistic connect.

To illustrate both the limitations of the 'abstract representation' view and the potential of a more process-based approach, we turn to an ongoing debate on spatial prepositions. Within this domain, one group of researchers has claimed that people use overlapping representations in linguistic and non-linguistic tasks, while a second group has claimed that different representations are used in these two types of task. Importantly, both sets of claims focus on representations in the abstract. We then sketch a different view by applying our neurally inspired model of spatial working memory—the Dynamic Field Theory (DFT)—to the issue of how people activate and use spatial information in linguistic tasks. Our analysis suggests that linguistic and non-linguistic behavior can arise from a single, integrated system that has a representational format different from what previous researchers have claimed. We then test some novel implications of our model to highlight the fact that a process-based view offers new ways to probe the linguistic/non-linguistic connection.

## 6.4  Missing the connection: the challenges of focusing on representation in the abstract

*6.4.1  Hayward & Tarr (1995): shared linguistic and perceptual representations of space*

To explore the possible connections between linguistic and sensorimotor representations of space, Hayward & Tarr (1995) examined how object relations are linguistically and visually encoded. Participants were presented with a visual scene depicting a referent object and a target object that appeared in varying locations. Participants were asked to generate a preposition describing the relationship. Results suggested that the prototypical spatial positions for 'above' and 'below' lie along a vertical reference axis, and prototypical spatial positions for 'left' and 'right' lie along a horizontal axis. In addition, use of these terms declined as target positions deviated from the respective axes.

Next, Hayward & Tarr extended these findings by using a preposition ratings task. In the ratings task, participants were asked to rate on a scale of 1 (least applicable) to 7 (most applicable) the applicability of a given spatial term (e.g. 'above') to a relationship between two objects. This ratings task is particularly valuable because it permits more graded quantification and metric manipulation of linguistic representations beyond the standard gross linguistic output (e.g. 'above'/'not above'). It therefore provides a means of empirically bridging the gap between metric, dynamic sensorimotor representations and discrete linguistic representations. Results from this ratings task showed strong metric effects of spatial language use around the vertical and horizontal axes. For instance, 'above' ratings were highest along the vertical axis and systematically decreased as the target object's position deviated from the vertical axis. Hayward & Tarr concluded that this ratings gradient across spatial positions reflected the use of prototypical vertical and horizontal reference axes.

To compare the representational prototypes of spatial language with visual representations of space, Hayward & Tarr examined performance on location memory and same-different discrimination tasks. Importantly, they found that the areas of highest spatial recall accuracy were aligned with the reference axes used as prototypes in the ratings task. Performance in the same/different location task yielded similar findings, showing that discrimination was best along the vertical and horizontal axes. Collectively, data from these four experiments point to a shared representational spatial structure between linguistic and sensorimotor systems with spatial prototypes along the cardinal axes. Such prototypes lead to high linguistic ratings and a high degree of accuracy in sensorimotor tasks for targets aligned with the axes.

## 6.4.2 *Crawford, Regier, and Huttenlocher (2000): distinct linguistic and perceptual representations of space*

Results from Crawford et al. (2000) present a different picture. Like Hayward & Tarr, these researchers probed both linguistic and non-linguistic representations of space by analyzing 'above' ratings as well as spatial memory performance. Results showed an 'above' ratings gradient aligned with the vertical axis similar to that of Hayward & Tarr (1995). Counter to the claims of representational similarity, however, Crawford et al. also found that location memory responses were biased *away* from the vertical axis when participants had to recall the locations of targets to the left and right of this axis. To account for these data, Crawford and colleagues proposed that the cardinal axes function as *prototypes* in the linguistic task (see Figure 6.1a) but serve as *category boundaries* in the spatial memory task (Figure 6.1b). Moreover, the diagonal

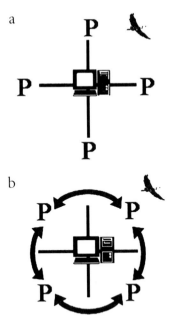

FIGURE 6.1. (a) Proposed layout of spatial prototypes (P) relative to a reference object (computer) and a target object (bird) in the linguistic task from Hayward & Tarr (1995) and Crawford et al. (2000). According to Hayward & Tarr, the same figure captures spatial prototypes in the non-linguistic task. (b) Proposed layout of spatial prototypes in non-linguistic tasks according to Crawford et al. Arrows in (b) indicate direction of bias in the spatial recall task. Lines in (b) indicate location of category boundaries.

axes in the task space, while serving no particular function in the linguistic task, serve as prototypes for spatial memory (Figure 6.1b) (Engebretson & Huttenlocher 1996; Huttenlocher, Hedges, & Duncan 1991). Thus, while both linguistic and non-linguistic spatial representations use the cardinal axes, these axes serve functionally distinct representational roles in the two tasks. It appears, therefore, that linguistic and non-linguistic representations of space differ in critical ways.

### 6.4.3 *A prototypical debate*

Results of the studies described above suggest that the cardinal axes serve as prototypical locations for spatial prepositions like 'above'. At issue, however, is what accounts for performance in the non-linguistic tasks—prototypes along the cardinal axes (Figure 6.1a) or prototypes along the diagonals (Figure 6.1b)? Both sets of researchers present standard evidence of prototype effects— graded performance around some special spatial locations. The challenge is that there appear to be two sets of special locations. Specifically, recall accuracy is highest when targets are near the cardinal axes and declines systematically as the target object is moved away from the axes, while at the same time bias is largest near the cardinal axes, declining systematically as one moves closer to the diagonal axes. Given these two sets of special locations—near the cardinal axes and near the diagonal axes—how do we know which layout of prototypes is correct?

Crawford et al. seem to present a compelling case by focusing on a critical issue: what goes into making a recall response in these tasks? In particular, Crawford and colleagues used their Category Adjustment (CA) model to explain why adults' responses are biased away from cardinal axes and toward the diagonals. According to this model, people encode two types of spatial information in recall tasks: *fine-grained* information about the target location (e.g. angular deviation) and the region or *category* in which the target is located. Data from a variety of studies suggest that adults tend to subdivide space using vertical and horizontal axes (Engebretson & Huttenlocher 1996; Huttenlocher et al. 1991; Nelson & Chaiklin 1980). This places prototypes at the centers of these regions, that is, along the diagonals of the task space. At recall, fine-grained and categorical information are combined to produce a response. Importantly, these two types of information can be weighted differently. If, for example, fine-grained information is uncertain (as is the case after short-term delays), categorical information can be weighted more heavily, resulting in a bias toward the prototype of the category. This accounts for the bias toward the diagonals in Crawford et al. (2000). It also accounts for the improved accuracy along the cardinal axes, because recall of targets aligned with a category boundary can be quite accurate (see Huttenlocher et al. 1991).

Given that Crawford et al. grounded their account of spatial memory biases in a formal theory of spatial recall that does not use prototypes along the cardinal axes, it appears that there are important differences in the representations underlying linguistic and non-linguistic performance. However, there are two limitations to this story. The first has to do with constraints provided by the CA model. Although this model can explain the biases that arise in recall tasks once one has specified the location of category boundaries, prototypes, and the certainty of fine-grained and categorical information, we do not know the processes that specify these things. That is, we do not know the factors that determine where category boundaries should go, what factors influence the certainty of spatial information, and so on. More recent work has documented some of these factors (Hund, Plumert, & Benney 2002; Plumert & Hund 2001; Spencer & Hund 2003; Spencer et al. 2007) but these details are not specified *a priori* by the CA model (for a recent modification of the CA model in this direction, see Hund & Plumert 2002; 2003).

Why are these details important? In the context of the linguistic/non-linguistic debate, this issue is central because both spatial language and spatial memory use the cardinal axes in some way. Specifying precisely what these axes do in both cases and how these axes are linked up to the representational states in question is critical if we are to evaluate the different claims. Put differently, we contend that it is important to specify the process that links the sensorimotor (e.g. perception of the cardinal and diagonal symmetry axes) and the cognitive (e.g. spatial prototypes). Note that this critique of the CA model does not indicate that this model is incorrect. Rather, we think the time is ripe to move the ideas captured by this model to the next level, that is, to the level of process.

A second limitation of the Crawford et al. story is that it fails to specify what is happening on the linguistic side: neither Crawford et al. nor Hayward and Tarr provided a formalized theory of spatial language performance. A recent model proposed by Regier and Carlson (2001)—the AVS model—specifies how prototypicality effects might arise in ratings tasks. Interestingly, this model can account for prototypicality effects without using prototypes *per se*. Rather, this model scales ratings by the difference between an attentionally weighted vector from the reference object to the target object and the cardinal axes in question (e.g. the vertical axis in the case of 'above' ratings). Thus, this model moves closer to explaining how ratings performance arises from processes that link the cardinal axes to representations of the target location. Unfortunately, this model says nothing about spatial recall performance. As such, it is not possible to directly compare the CA account of spatial memory biases and the AVS account of ratings performance.

In the sections that follow, we describe a new theory of spatial working memory that we contend can overcome both limitations described above. In particular, this model overcomes the limitation of the CA model by specifying how perception of symmetry axes is linked to the representational states associated with target locations in spatial recall tasks. The critical insight here is that we can account for both accuracy along the cardinal axes and bias away from the cardinal axes without postulating category boundaries and prototypes; rather, such effects arise due to the coupling between perception of reference axes in the task space—visible edges and axes of symmetry—with working memory processes that serve to actively maintain location information. With regard to the second limitation—the absence of a formal model of both spatial recall and spatial preposition use—we sketch an extension of our model that can account for prototypicality effects in linguistic ratings tasks. Although this extension requires further development (which we point toward in the conclusions section), it is generative enough at present to produce novel predictions which we test empirically.

## 6.5  A process approach to the linguistic/non-linguistic connection

### 6.5.1  *The Dynamic Field Theory: a process account of spatial working memory*

Data from Hayward and Tarr (1995) and Crawford et al. (2000) point toward two types of prototypicality effects in spatial memory—higher accuracy and greater bias near cardinal axes. Although the CA model explains these biases using two types of representation—boundaries and prototypes—our Dynamic Field Theory (DFT) suggests that both effects actually arise from the interaction of perceived reference frames and information actively maintained in spatial working memory (Simmering, Schutte, & Schöner 2008; Spencer & Schöner 2003; Spencer et al. 2007). That is, the DFT provides a formalized process account of spatial memory bias away from reference axes *without positing prototypes*.

The DFT is a dynamic systems approach to spatial cognition instantiated in a particular type of neural network called a dynamic neural field. The DFT accounts for the spatial recall performance of younger children (2–3 years), older children (6 and 11 years), and adults (see Spencer et al. 2007). A simulation of the DFT performing a single spatial recall trial is shown in Plate 1. The model is made up of several layers (or fields) of neurons. In each layer, the neurons are lined up along the x-axis according to their 'preferred' location, that is, the location in space that produces maximal activation of each neuron. The activation of each neuron is plotted along the y-axis, and time is on the z-axis. The top layer in each panel is the perceptual field, PF. This field captures perceived events in the task space, such as the appearance of a target, as well as any stable perceptual cues in the task space, such as the midline symmetry axis probed in many studies of spatial recall. This layer sends excitation to both of the other layers (see green arrows). The third layer, SWM, is the working memory field. This field receives weak input from perceived events in the task space and stronger input from the perceptual field. The SWM field is primarily responsible for maintaining a memory of the target location through self-sustaining activation—a neurally plausible mechanism for the maintenance of task-relevant information in populations of neurons (Amari 1989; Amari & Arbib 1977; Compte, Brunel, Goldman-Rakic, & Wang 2000; Trappenberg, Dorris, Munoz, & Klein 2001). The second layer, Inhib, is an inhibitory layer that receives input from and projects inhibition broadly back to both the perceptual field and the working memory field. Note that the layered structure shown in Figure 6.2 was inspired by the cytoarchitecture of visual cortex (see Douglas & Martin 1998). Note also that our full theory of spatial cognition includes longer-term memory layers that we will not consider here because they do not affect the hypotheses tested below (for an overview of the full model, see Spencer et al. 2007).

The working memory field, SWM, is able to maintain an activation pattern because of the way the neurons interact with each other. Specifically, neurons that are activated excite neurons that code for locations that are close by, and—through the Inhib layer—inhibit neurons that code for locations that are far away. The result is an emergent form of local excitation/lateral inhibition which sustains activation in working memory in the absence of inputs from the perceptual layer (see Amari 1989; Amari & Arbib 1977; Compte et al. 2000 for neural network models that use similar dynamics).

Considered together, the layers in Plate 1 capture the real-time processes that underlie performance on a single spatial recall trial. At the start of the trial, there is activation in the perceptual field associated with perceived reference axes in the task space (see reference input arrow in Plate 1a), for instance, visible edges and axes of symmetry (Palmer & Hemenway 1978; Wenderoth & van der Zwan 1991). This is a weak input and is not strong enough to generate a self-sustaining peak in the SWM field, though it does create an activation peak in perceptual field. Next, the target turns on and creates a strong peak in PF which drives up activation at associated sites in the SWM field (see target input arrow in Plate 1a). When the target turns off, the target activation in PF dies out, but the target-related peak of activation remains stable in SWM. In addition, activation associated with the reference axis continues to influence the PF because the reference axis is supported by readily available perceptual cues (see peak in PF during the delay).

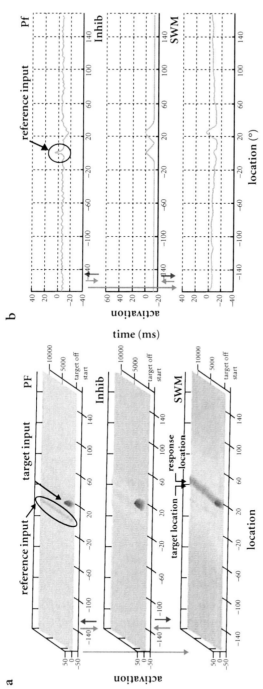

1. A simulation of the Dynamic Field Theory performing a single spatial recall trial. Panels in (a) represent: perceptual field [PF]; inhibitory field [Inhib]; spatial working memory field [SWM]. Arrows represent interaction between fields. Green arrows represent excitatory connections and red arrows represent inhibitory connections. In each field, location is represented along the x-axis (with midline at location o), activation along the y-axis, and time along the z-axis. The trial begins at the front of the figure and moves toward the back. (b) Time slices through PF, Inhib, and SWM at the end of the delay for the model shown in (a). See text for additional details. For improved image quality and colour representation see Plate 1.

Central to the recall biases reported by Huttenlocher and colleagues (1991) is how reference-related perceptual input affects neurons in the working memory field during the delay. Figure 6.2b shows a time slice of the SWM field at the end of the delay. As can be seen in the figure, the working memory peak has slightly lower activation on the left side. This lower activation is due to the strong inhibition around midline created by the reference-related peak in PF (see circle in Plate 1b). The greater inhibition on the left side of the peak in SWM effectively 'pushes' the peak away from midline during the delay. Note that working memory peaks are not always dominated by inhibition as in Figure 6.2b. For instance, if the working memory peak were positioned very close to or aligned with midline (location 0), it would be either attracted toward or stabilized by the excitatory reference input. This explains why spatial recall performance is quite accurate for targets aligned with the cardinal axes (for related results, see Engebretson & Huttenlocher 1996; Hund & Spencer 2003; Spencer & Hund 2003).

In summary, the DFT provides a process-based alternative to the CA model. Critically, the DFT links spatial memory biases to a process that integrates remembered information in working memory with perceived reference frames—the cardinal axes of the task space—the same reference frames implicated in linguistic performance. As a result, Crawford et al.'s central argument against Hayward & Tarr's claim of shared structure between linguistic and non-linguistic representations of space—that memory is biased away from a category boundary—no longer follows obligatorily from the data. This provides the impetus to once again consider the possibility that there is a direct link between spatial memory and spatial language.

### 6.5.2 *Connecting the Dynamic Field Theory and spatial language*

Given that we have proposed a process-based account of the link between cardinal axes and spatial memory, we can ask whether this proposed link between the sensorimotor and the cognitive can be extended to the case of spatial language. The central issue this raises is: what is the connection between the representational states associated with space captured by our theory and the representational states underlying words? A simple way to conceptualize this link is depicted in Figure 6.2, which captures the use of the spatial preposition 'above' to describe a target presented at $-20°$. This figure shows the working memory field depicted in Figure 6.2b reciprocally coupled to a linguistic node that represents the label 'above'. The $-20°$ target location is captured by the Mexican-hat-shaped activation distribution which arises from the locally excitatory interactions among neurons in the SWM layer and lateral inhibition from activation in the Inhib layer. The forward projection from SWM to the 'above' node is spatially structured by systematically varying the connection strengths (captured by the Gaussian distribution of connection lengths) around the vertical axis. In particular, neurons in SWM associated with the vertical axis (location 0) project activation most strongly onto the 'above' node, while neurons to the far left and right of the excitatory field project activation quite weakly onto this node. These variations in synaptic strength are meant to reflect the long-term statistical probabilities of spatial preposition use. In particular, we hypothesize that over the course of development, 'above' is used most often when referring to cases where a target object is close to a vertical axis and less often when a target object is to the far left and right of a vertical axis. This is consistent with findings from Hayward & Tarr (1995) showing that spontaneous use of prepositions like 'above' and 'over' declines as target objects diverge systematically from a vertical or 'midline' axis (see also Franklin & Henkel 1995). Note that the strength of the projection gradient depicted in Figure 6.2 is somewhat arbitrary: the gradient does not have to be very strong for our account of spatial language performance to work (see below). Note also that we only consider the forward projection from SWM to the 'above' node in this chapter. We view the coupling between spatial memory and spatial language as reciprocal in nature; thus, the vectors in Figure 6.2 go in both directions. The details of this reciprocal coupling, however, are beyond the scope of the present chapter (see Lipinski, Spencer, & Samuelson 2009b for an empirical probe of the reciprocal nature of these connections).

How can the model depicted in Figure 6.2 be applied to capture performance in spatial language tasks? Essentially, this model provides an account for why some locations might be perceived to be better examples of 'above' than others. In particular, a target-related peak of activation in SWM close to the vertical axis (e.g. the activation peak in Figure 6.2 to the left of location 0) would strongly activate the 'above' node. By contrast, a target-related peak in SWM far from the axis would weakly activate the 'above' node. We turn these activations into a linguistic 'above' rating by scaling the amount of activation to the magnitude of the rating. Concretely, ratings should be highest when targets are aligned with the vertical axis, and should fall off systematically as peaks of activation in SWM are shifted to the left or right. This is similar to the approach adopted by Regier & Carlson's AVS model (2001). Although this account of ratings performance is, admittedly, simplistic, we contend that it has a clear strength: it grounds linguistic performance in the real-time events that occur in spatial language tasks, and places primary emphasis on the real-time activation of lexical representational states that are reciprocally coupled to spatial working memory. In the next section, we empirically demonstrate

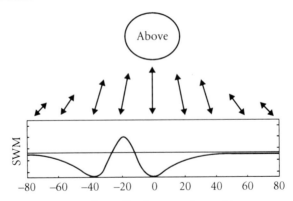

FIGURE 6.2. Proposed reciprocal coupling between the working memory field in Plate 1 and a linguistic node representing the label 'above'. The projections between SWM and this node are spatially structured by systematically varying the connection strengths (captured here by the Gaussian distribution of connection lengths) around the vertical axis (location 0). The $-20°$ target location is captured by the Mexican-hat-shaped activation distribution. See text for additional details.

that this emphasis on process can shed new light on what is happening in spatial language tasks.

## 6.6  An empirical test of the DFT approach to spatial language

Inspired by the model sketched in Figure 6.2, we recently conducted a study designed to investigate whether linguistic and non-linguistic processes are temporally connected in spatial tasks. In particular, we asked whether the processes that create delay-dependent spatial drift in spatial working memory might also leave empirical signatures in a spatial language task. Toward this end, we used the ratings task from Hayward & Tarr (1995), given its capacity to reveal quantifiable metric effects and its centrality in the spatial language literature (e.g. Crawford et al. 2000; Hayward & Tarr 1995; Logan & Sadler 1996; Regier & Carlson 2001). We predicted that if spatial language and spatial memory are coupled together as shown in Figure 6.2, then 'above' ratings should become systematically lower for targets to the left and right of the vertical axis as memory delays increase—that is, the 'above' node should become systematically less active as peaks of activation in SWM drift away from the vertical axis. Furthermore, the variability of ratings performance should increase over delays and be systematically lower when participants rate targets aligned with the cardinal axes. These predictions regarding response variability mirror effects we have reported in our previous studies of spatial recall (e.g. Spencer & Hund 2002).

### 6.6.1 *Design and methods*

To test this prediction, we used a variant of the basic 'spaceship' task used in our previous spatial working memory studies (e.g. Schutte & Spencer 2002; Spencer & Hund 2002; 2003). Participants were seated at a large (0.921 × 1.194 m), opaque, homogeneous tabletop. Experimental sessions were conducted in a dimly lit room with black curtains covering all external landmarks. In addition, a curved border was added to occlude the corners of the table, thereby occluding the diagonal symmetry axes. Thus, visible reference cues included the edges of the table and its axes of symmetry as well as the objects included in our visual displays (see below).

On each trial, a single reference disk appeared along the midline (i.e. 'vertical') symmetry axis, 30 cm in front of the participant. This disk remained visible throughout each trial. Next, the participant moved a computer mouse on top of this disk and a random number between 100 and 500 appeared in the center of the table. Participants were instructed to count backwards by 1's from this number until the computer prompted them to make a response. This counting task occupied verbal working memory, preventing participants from verbally encoding and maintaining the position of the spaceship on trials with a memory delay. This was important because we wanted to examine whether verbal performance would show evidence of delay-dependent 'drift'. This also took care of a potentially important experimental confound in Hayward & Tarr (1995) and Crawford et al. (2000). In both of these studies, the verbal responses could be formulated while the target was visible; spatial recall responses, on the other hand, were given after a memory delay. Thus, any differences between spatial language and spatial memory performance might be simply due to processing in the absence of a memory delay in one task and processing following a delay in the other.

After participants started counting, a small, spaceship-shaped target appeared on the table for 2 sec. Next, participants gave a response based on one of two prompts spoken by the computer. For *spatial memory trials*, participants moved the mouse cursor to the remembered target location when the computer said 'Ready-Set-Go'. For *spatial language rating trials*, participants gave a verbal rating when the computer said 'Please give your "Above" rating'. The computer prompts were both 1,500 msec. in duration. On ratings trials, participants rated on a scale of 1 ('definitely not above') to 9 ('definitely above') the extent to which the sentence 'The ship is ABOVE the dot' described the spaceship's location relative to the reference disk. Ratings and recall trials were randomly intermixed, and responses were generated following a 0 sec. or 10 sec. delay. In particular, in the No Delay condition, the end of the computer prompt coincided with the disappearance of the target, while in the Delay condition, the prompt ended 10 sec. after the disappearance of the target. Targets appeared at a constant radius of 15 cm relative to the reference disk and at 19 different locations relative to the midline axis (0°): every 10° from −70° to +70° as well as ±90° and ±110°.

### 6.6.2 *Results and discussion*

Figure 6.3a shows mean directional errors on the memory trials across target locations and delays. Positive errors indicate clockwise errors relative to midline (vertical), while negative errors indicate counterclockwise errors. As can be seen in the figure, participants' responses were quite accurate in the No Delay condition. After 10 sec., however, responses to targets to the left and right of midline were systematically biased away from this axis (see also Spencer & Hund 2002). This bias gradually increased and then decreased as targets moved away from midline, reducing considerably at the horizontal or left-to-right axis (i.e. ±90°). These data were analyzed in an ANOVA with Target and Delay as within-subject factors. This analysis revealed a significant main effect of Target, $F(18 234) = 20.6, p < .001$, as well as a significant Delay by Target interaction, $F(18 234) = 19.4, p < .001$. This interaction is clearly evident in Figure 6.3a.

Similar results were obtained in analyses of response variability (standard deviations of performance to each target at each delay; see Figure 6.3b). There were significant main effects of Delay, $F(1 13) = 172.3, p < .001$, and Target, $F(18 234) = 5.4$, $p < .001$, as well as a significant Delay by Target interaction, $F(18 234) = 3.4, p < .001$. As can be seen in Figure 6.3b, variability was higher in the 10 sec. delay condition, and responses to targets to the left and right of midline were more variable than responses to the targets aligned with the cardinal axes. These results are consistent with predictions of the DFT that memory for locations aligned with symmetry axes is more stable than memories for targets that show delay-dependent drift.

The first critical question was whether delay-dependent spatial drift would be evident in participants' ratings performance. Figure 6.4 shows that this was indeed the case. Overall, 'above' ratings in the spaceship task followed a gradient similar to that obtained by Hayward & Tarr (1995) and Crawford et al. (2000); however, ratings were systematically lower for targets to the left and right of midline after the delay (see Figure 6.4a). An ANOVA on these ratings data with Target and Delay as within-subjects factors revealed a significant main effect of Target, $F(18, 234) = 240.2, p < .001$. More importantly, there was a significant decrease in ratings over Delay, $F(1, 13) = 12.5, p = .004$, as well as a trend toward a Delay

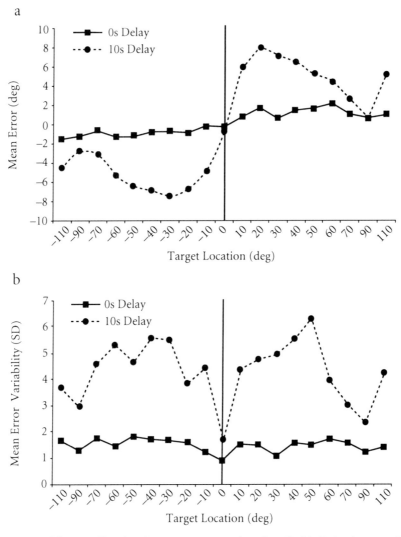

FIGURE 6.3. (a) Mean directional error across target locations for No Delay (o sec.; solid line) and Delay (10 sec.; dashed line) location memory trials. Positive errors indicate clockwise errors and negative errors indicate counter-clockwise errors. (b) Mean error variability (SDs) for No Delay (o sec.; solid line) and Delay (10 sec.; dashed line) location memory trials. Solid vertical line in each panel marks the midline of the task space.

by Target interaction, $F(18\ 234) = 1.5$, $p < .10$. This systematic decrease in ratings responses as a function of delay—particularly for targets to the left and right of the reference axis—is consistent with the proposal that there is a shared representational process used in both the spatial memory and spatial language tasks.

a

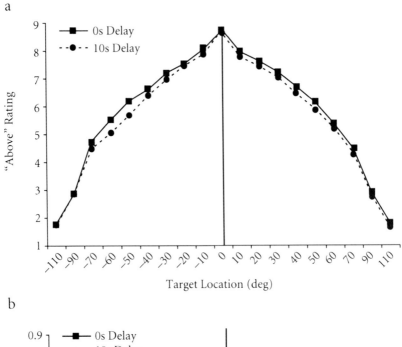

b

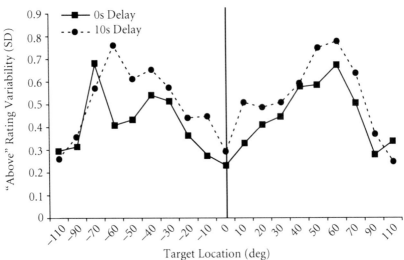

FIGURE 6.4. (a) Mean 'Above' ratings across target locations for No Delay (0 sec.; solid line) and Delay (10 sec.; dashed line) trials where '9' indicates the target is 'definitely above' the reference dot and '1' indicates the target is 'definitely NOT above' the reference dot. (b) Mean 'Above' ratings variability (SDs) for No Delay (0 sec.; solid line) and Delay (10 sec.; dashed line) trials. Solid vertical line in each panel marks the midline of the task space.

Given the effects of delay on response variability *and* bias in the spatial memory task, a second critical question is whether such variability effects would also emerge in our analyses of ratings performance. If ratings performance failed to reflect the same pattern of variability as that established in spatial memory (namely, lower variability for targets appearing along the vertical axis), it would indicate some difference in the underlying representational processes required for the linguistic and non-linguistic spatial tasks. If, on the other hand, the same general pattern is obtained, it bolsters our claim that both tasks rely on the same underlying representational process. Our analyses of ratings variability were consistent with the latter, showing significant main effects of both Target, $F(18\ 234) = 3.4, p < .001$, and Delay, $F(1\ 13) = 8.8, p = .01$. As can be seen in Figure 6.4b, the variability in ratings performance was lower for targets aligned with the cardinal axes, and systematically increased as targets moved away from midline. Moreover, variability increased systematically over delay. These findings are similar to results obtained in the spatial memory task.

Overall, the similar effects of delay and target for the spatial memory and spatial language tasks points toward a shared representational process for both tasks. However, in contrast to the large delay effects in the spatial memory task (see Figure 6.3a), the effect of delay on ratings means in Figure 6.4a appears small. Given this, it is important to ask whether the significant delay effect in the ratings task is, in fact, a *meaningful* effect. To address this question, we compared spatial memory and ratings responses directly by converting the ratings 'drift' apparent in Figure 6.4a into a spatial deviation measure. In particular, for each target within the range ±60°,[1] we converted the ratings data in a two-step process. To illustrate this process, consider how we converted the data for the +10° target, the 'anchor' location in this example. First, we took the change in ratings in the No Delay condition between the anchor (10°) and the adjacent target moving away from midline (i.e. 20°) and divided this change by 10°—the separation between adjacent targets. This indicated the amount participants changed their rating in our baseline condition (i.e. No Delay) as we moved the anchor target 10° further from midline. Second, we scaled the change in rating over delay for the anchor target by this No Delay deviation measure (e.g. conversion score for the 10° target.= (change in 10 s delay rating at 10°) ˙ 10° / (change in 0 s delay rating between 10° and 20°) ).

The converted ratings data for all targets within the ±60° range are plotted in conjunction with the recall data in Figure 6.5. If the drift underlying

---

[1] For targets greater than 70° away from midline, adjacent targets were 20° apart. Given this change in spatial separation, we only converted the ratings data from targets ±60° from midline.

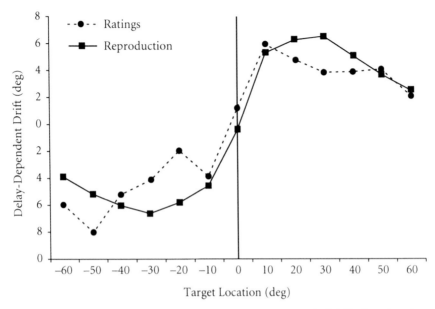

FIGURE 6.5. Comparison between location memory errors (solid line) and ratings drift converted to degrees (dashed line). Solid vertical line marks the midline of the task space. See text for details of ratings conversion method.

performance in the ratings task is produced by the same process that creates drift in the recall task, then these data should line up. Although differences in performance across tasks do exist, the converted ratings data show remarkable overlap with the recall data across target locations. This provides strong initial support for the prediction we generated from the modified dynamic field model shown in Figure 6.2, suggesting that a shared working memory process underlies performance in both tasks.

## 6.7 Conclusions

Understanding the relationship between linguistic and non-linguistic systems is a critical issue within cognitive science. Spatial language is of central importance here because it is an unambiguous case of these putatively different systems coming together. Although recent efforts have advanced our understanding of the link between spatial language and memory, we have argued in this chapter that previous approaches are limited in two related ways: these approaches have focused too narrowly on static representation and have led to under-constrained theories. To illustrate an alternative approach, we presented an overview of our dynamic field theory of spatial working memory and applied this process model to the use of spatial prepositions. Moreover, we presented preliminary empirical findings that supported a novel prediction of this model—that linguistic ratings would show signatures of delay-dependent 'drift' in both changes in mean ratings over delay and response variability. We contend that these results demonstrate the utility of our approach and suggest that sensorimotor and linguistic systems are intricately linked. This supports the view presented by Hayward & Tarr (1995) and others (see also Barsalou 1999; Richardson et al. 2003; Spivey-Knowlton et al. 1998; Zwaan et al. 2004) that sensorimotor and linguistic representations overlap. Importantly, however, it builds on this perspective by grounding claims about representation in a formal model that specifies the time-dependent processes linking perception of reference frames to representational states in working memory.

Although the model and data we present in this chapter support our claim that process-based approaches can shed new light on the linguistic/non-linguistic connection, these are only first steps. Clearly, there is much more theoretical and empirical work to do to demonstrate that our approach can move beyond previous accounts toward a more theoretically constrained future. In this spirit, the sections below address three questions: what have we accomplished, what remains to be accomplished, and how does our model fit with other related models in the spatial memory and spatial language literatures?

### 6.7.1 *The DFT and spatial language: what have we accomplished?*

The model and data presented in this chapter are firmly positioned between arguments by Hayward & Tarr (1995) and Crawford et al. (2000). On one hand, our data show a time-dependent link between spatial language and spatial memory, consistent with the claim by Hayward & Tarr (1995) that linguistic and non-linguistic representations have considerable overlap within the spatial domain. On the other hand, our work also resonates with the move toward formal models by Crawford et al. (2000). In particular, our modeling work emerged from a focus on the question originally addressed by the Category Adjustment model: what goes into making a recall response (Huttenlocher et al. 1991)? By focusing on the processes that link cardinal axes to representational states in spatial working memory, the DFT provides a new answer to this question that does not have recourse to spatial prototypes. The absence of spatial prototypes in our model allowed us to reconsider the link between performance in spatial recall and ratings tasks. We proposed a new view that directly couples SWM and the activation of label nodes representing spatial terms like 'above'. This new view moves beyond past approaches in two key ways: (1) it grounds both recall and ratings performance in time-dependent perceptual and working memory processes, and (2) it provides a formal account of how people generate *both* types of responses.

Importantly, we also demonstrated in this chapter that the dynamic field approach is empirically productive. We generated a set of novel predictions that ratings of targets to the left and right of midline would be lower after a short-term delay, and that response variability in the ratings task would increase over delays and be lower for targets aligned with the cardinal axes. Analyses of both mean ratings and response variability were consistent with these predictions.

These results are not trivial because we predicted *lower* ratings over delay when other views appear to predict *higher* ratings. In the CA model, for example, people rely more on spatial prototypes after short-term delays. If the spatial prototypes for language lie along the cardinal axes, as both Hayward & Tarr (1995) and Crawford et al. (2000) contend, ratings should have drifted toward these prototypes over delay—that is, people should have rated a target close to midline as a better example of 'above' after a delay relative to the No Delay condition. As predicted by the DFT, however, we found the opposite result. Indeed, the converted ratings data showed a high degree of overlap with spatial recall biases, suggesting that a shared process generated both types of response.

This discussion makes it clear that our model did, in fact, generate a novel prediction. But wouldn't *any* model in which sensorimotor and linguistic representations use the same underlying process make this prediction? We contend that the answer is 'no' because we predicted an entire suite of effects: a decrease in ratings over delays for targets to the left and right of the vertical axis; an increase in ratings response variability over delays; and lower ratings variability for targets aligned with the cardinal axes. It is important to note in this regard that our model provides a process-based account for both mean biases *and* response variability (see Schutte, Spencer, & Schöner 2003; Schutte & Spencer in press). This is rarely the case for models of spatial memory. For comparison, the CA model has not been used in the spatial domain to make predictions about response variability (although for a model that moves in a related direction see Huttenlocher, Hedges, & Vevea 2000).

Results showing signatures of delay-dependent spatial drift in both memory and ratings tasks are consistent with our predictions, but might these results be an artifact of how we structured the tasks? For instance, did we create an artificial link between spatial memory and spatial language by randomly intermixing recall and ratings trials? Perhaps in the face of this response uncertainty, participants prepared two responses in the delay conditions. This might have caused the two prepared responses to interact during the delay, leading to shared bias and shared response variability in the two tasks. Recent data suggest that this is not the case. We conducted a second version of the experiment reported here with recall and ratings trials split across two sessions (Lipinski, Spencer, & Samuelson 2009a). The key result comes from the condition where participants did the ratings task in session 1 and the recall task in session 2. Critically, these participants had no knowledge of the recall task during their first session. Results replicated the findings reported in this chapter.

A related concern is whether we created an artificial link between spatial memory and spatial language by preventing participants from making a rating when both the target and reference object were visible. Recall that this was not the case in Hayward & Tarr (1995) and Crawford et al. (2000): in these studies, ratings could be prepared when the target was visible. In our task, therefore, people had to make a rating using their memory of the target location, in some sense forcing participants to link spatial memory and language. We certainly agree that the nature of our task requires that people use their memory of the target location in the ratings task. Importantly, however, the model we sketched in Figure 6.3 accounts for performance both with and without an imposed memory delay. More specifically, this model would generate a systematic shift in ratings of 'above' as visible targets were moved away from the vertical axis, and it would generate accurate pointing movements to visible target locations. Thus, even if we did create an artificial link between memory and language in our experiment, the model we proposed is still useful because it suggests how performance in multiple task contexts can be seamlessly woven together within a single framework. Moreover, we claim that, although our ratings task is certainly artificial, the processes at work in our 'delay' tasks are not. In particular, there are many naturalistic situations where we need to use our memory of objects' locations to generate spatial descriptions. Indeed, it is possible that spatial prepositions are used *more frequently* in cases where the objects in question are not visible. When two people are staring at the same visible objects, verbal communication is simple: 'hand me that' along with a pointing gesture will suffice. By contrast, when objects are not visible, 'hand me that' no longer works. In these situations, spatial prepositions are critical to effective communication.

### 6.7.2 *The DFT and spatial language: what still needs to be accomplished?*

Although the dynamic field model we sketched in this chapter provides a solid first step in a process-based direction, it is clearly overly simplistic. Nevertheless, the structure of the model provides useful constraints as we look to the future. In particular, we see five challenges that must be addressed within this theoretical framework. First, we must specify the process that aligns labels with particular reference locations in SWM. In Figure 6.3, we 'manually' aligned the 'above' node with location 0 in SWM. The challenge is that adults can do this quite flexibly. Consider, for instance, what adults had to do in our task—they made 'above' ratings for targets presented in the horizontal plane. Although such judgements are not typical, participants had little difficulty adjusting to the task, and our results replicated the ratings gradient from studies that used a vertically oriented computer screen (Crawford et al. 2000; Hayward & Tarr 1995; Logan & Sadler 1996). The question is: what process accomplishes this flexible alignment? In our current model, we have an alignment process that matches perceived and remembered reference frames via a type of spatial correlation (Spencer et al. 2007). It is an open question, however, whether a related type of alignment process could work for the case of labels (for a robotic demonstration of this possibility, see Lipinski, Sandamirskaya, & Schöner 2009).

Next, we need to specify the process that structures the projection from SWM to the 'above' node. Conceptually, this gradient reflects the statistics of 'above' usage over development, but we need to specify the process that accumulates this statistical information. In past work, we have used activation in long-term memory fields to accumulate a type of statistical information across trials (Schutte & Spencer 2007; Simmering, Schutte, & Spencer 2008; Spencer et al. 2007; Thelen et al. 2001). Such long-term memory fields implement a form of Hebbian learning. A related issue is how to accumulate information across contexts. For instance, when young children are first learning the semantics of 'above', what process integrates use of this term across the diversity of situations in which this term is used? Put differently, what process accounts for generalization across contexts?

A third central component of our dynamic field model that needs further development is the nature of the bi-directional coupling between SWM and the 'above' node. Conceptually, coupling means that the establishment of stable patterns of activation within one layer should contribute to stable patterns in the other. Similarly, instability and drift within one layer should contribute to instability and drift within the other layer. The data presented in this chapter are consistent with the proposed link from SWM to the 'above' node, but what about coupling in the other direction? Recent experiments have confirmed that activation of a spatial term can stabilize spatial memory in some cases and amplify drift in others (Lipinski, Spencer, & Samuelson 2009b). Importantly, these results shed light on how the activation of labels projects back onto SWM and the situations in which this occurs.

The fourth challenge presented by our model is to expand beyond 'above' to handle multiple spatial prepositions. This requires that the processes we develop to handle the challenges above should generalize to other spatial labels. In this sense, we need to develop a formal, general theory of the link between space and words (see Lipinski, Sandamirskaya, & Schöner 2009). Furthermore, we need to expand the model to handle the labeling of locations with multiple spatial terms such as 'above and to the right' (see Franklin & Henkel 1995; Hayward & Tarr 1995). Such effects can be handled by neural connections among the nodes representing different labels; however, we must specify the process that structures these connections. In this context, it is useful to note that our treatment of spatial terms via the activation of individual label nodes is consistent with several recent models of categorization and category learning that treat labels as a single feature of objects (e.g. Love, Medin, & Gureckis 2004).

Consideration of multiple spatial prepositions leads to the final issue our approach must handle: the model must ultimately speak to issues central to language use, such as how the real-time processes of spatial memory and spatial language relate to symbolic capacities for syntax and type-token distinctions. These broader issues obviously present formidable challenges, but we contend that there is no easy way around such challenges if the goal is to provide a constrained, testable theory of the connection between linguistic and non-linguistic systems. Given the neurally inspired view proposed by Barsalou (1999), an intriguing possibility is that the dynamic field approach could offer a formal theoretical framework within which one could specify the details of a perceptual symbol system.

### 6.7.3  *Ties between our process-based approach and other models*

When discussing our dynamic field model, it is of course critical to consider alternative models that are moving in related process-based directions. Two models are relevant here. The first is Regier & Carlson's (2001) AVS model. This model incorporates the role of attention in the apprehension of spatial relations (Logan 1994; 1995) as well as the role of the geometric structure of the reference object (Regier & Carlson 2001). As mentioned previously, there is conceptual overlap between our dynamic field approach and AVS, in that both models scale ratings for prepositions like 'above' by the deviation between a reference axis and the target object. The manner by which these two models arrive at this deviation measure differs, however. In our model, this deviation is reflected in activation differences of the 'above' node that are structured by the projection gradient from SWM to this node. In AVS, by contrast, this deviation reflects the difference between a vertical axis and an attentionally weighted vector sum. A critical question for the future is whether these differences lead to divergent predictions. It is also important to note that AVS says nothing about performance in spatial recall tasks. As such, this model is not well positioned to examine links between spatial language and spatial memory.

A second related model is O'Keefe's (2003) Vector Grammar. This model is similar to AVS in that location vectors provide the link between the perceived structure of the environment and the use of spatial prepositions. In contrast to AVS, however, these vectors are derived from a model of place cell receptive field activations (see Hartley, Burgess, Lever, Cacucci, & O'Keefe 2000). The Vector Grammar approach shares conceptual overlap with the model we sketched in this chapter. In particular, both the place cell model by Hartley et al. (2000) and our dynamic field approach (Amari 1977; 1989; Amari & Arbib 1977; Bastian, Riehle, Erlhagen, & Schöner 1998; Bastian, Schöner, & Riehle 2003; Erlhagen, Bastian, Jancke, Riehle, & Schöner 1999) are grounded in neurophysiology. Moreover, there is a strong spatial memory component to O'Keefe's Vector Grammar approach in that it explicitly attempts to link 'Cognitive Maps' (Tolman 1948) with linguistic 'Narrative Maps' (O'Keefe 2003). Beyond these areas of overlap, it is not yet clear the extent to which linguistic and non-linguistic spatial representational states are truly coupled or simply analogous in the Vector Grammar model. It will be important to evaluate this linguistic/non-linguistic link in the future as both modeling frameworks are expanded.

### 6.7.4 *Summary: toward a more process-based future*

We end this chapter by reiterating three central themes. First, we contend that the linguistic/non-linguistic connection must remain a central focus in cognitive science. Although tackling this issue presents formidable challenges, we think that the time is ripe to revisit it afresh given recent advances in both empirical techniques—for instance, the eye-tracking methods pioneered by Tanenhaus and colleagues (Tanenhaus et al. 1995)—and formal theoretical approaches—for instance, the dynamic field framework presented here (e.g. Spencer et al. 2007). Second, although focusing on representations in the abstract appears to be a useful simplification of the linguistic/non-linguistic link, this approach is not a panacea. Instead, we contend that such efforts can lead to under-constrained theories, a point we illustrated using an example from the spatial preposition literature. Third, we think close ties between theory and experiment can move the spatial language literature toward a process-based and theoretically constrained future. A growing number of empirical studies have explored the real-time linkages between linguistic and non-linguistic systems (e.g. Richardson et al. 2003; Spivey-Knowlton et al. 1998; Tanenhaus et al. 1995). This exciting work provides an excellent foundation for the development of the formal, process-based approach we have sketched here. Clearly there is a long way to go in this regard, but efforts that link formal theory and empirical work in this domain are critical if we are to address one of the most vexing issues in cognitive science today—the connection between the sensorimotor and the linguistic.

**Author Notes**

We wish to thank Gregor Schöner for helpful discussions during the development of the model sketched here. This research was supported by NIMH (RO1 MH62480) and NSF (BCS 00-91757) awarded to John P. Spencer. Additional funding was provided by the Obermann Center for Advanced Studies at the University of Iowa.

# 7

## Tethering to the World, Coming Undone

BARBARA LANDAU, KIRSTEN O'HEARN, AND JAMES E. HOFFMAN

### 7.1 Introduction

Embodiment in spatial cognition. The very sound of it elicits a sense of mystery and vagueness that would appear inappropriate for scientific inquiry. Yet, as the chapters in this volume attest, the notion of embodiment—the idea that the body and its interactions with the world support, anchor, guide, and may even substitute for cognitive representations—has been gaining attention in cognitive science. The general idea of embodiment is this: Despite arguments in favor of abstract internal cognitive representations of space, our physical anchoring to the world has significant consequences for the way that we carry out spatial computations, for the efficiency and accuracy with which we do so, and perhaps even for how we come to develop spatial representations of the world.

Students of development will recognize a theme that was a deep part of Piaget's view that the sensory and motor activities of the infant are the core building blocks of abstract cognitive representations (Piaget 1954; Piaget & Inhelder 1956). But recent approaches suggest that the role of embodiment in cognition has consequences that persist long after childhood. Adults—like children—live in a 3-dimensional physical world, and thus necessarily connect to that world. We are connected through unconscious mechanisms such as eye movements that help us seek information in the world. We are connected by our posture, spending a great deal of waking time upright, viewing objects and layouts from a perspective in which gravity matters. And we are connected by our bodily movement, which is spatial by nature, and therefore provides a critical foundation for perceiving and remembering locations and producing new actions. These connections between our bodies and the external world help form and constrain our mental representations, providing links—or anchors—between entities in the world and their representations in our mind.

In this chapter, we ask about the specific role that these embodied connections play in theories of human spatial cognition. As Clark (1999) points out, one can contrast 'simple embodiment' with 'radical embodiment'. Proponents of *simple embodiment* use empirical facts about the importance of mechanisms such as eye movements, upright posture, the nature of terrain, etc. to better understand the nature of the internal representations that we use to know the world. For example, Ballard, Hayhoe, Pook, & Rao (1997) suggest that eye movements play the role of 'deictic pointers', which allow us to revisit the same location over time, discovering different object properties on each visit, and thereby binding them together with relatively little reliance on memory. In this approach, the embodied framework provides insight into the working mechanisms of visual and spatial cognition. Specifically, Ballard suggests that the visual system constructs spatial representations in a piecemeal fashion by relying on the world as an external memory or 'blackboard'. In contrast, *radical embodiment* seeks to show that internal representations are unnecessary—that one can explain many cognitive phenomena without notions such as abstract mental representation. For example, Thelen and Smith (1998) lay out a dynamical systems approach to walking and other actions that explains these in terms of systematic, continuous local interactions between body and world that do not require any role for mental representations. More radically, they extend the framework to higher level cognition—word learning, categorization, and the like.

Our view will be closer to simple embodiment. In particular, we will argue that, although interactions of the body and world play an interesting role in the development and use of rich spatial representations of the world, these interactions by themselves cannot be a substitute for abstract representations. Indeed, we will argue that real advances in spatial cognitive functions require that we become untethered from the physical world—capable of thought that goes beyond our current connections with the world. This kind of thought requires spatial representations that are rich, robust, and amenable to mental manipulation.

In making these arguments, we will use evidence from our recent studies of spatial representation in people with Williams syndrome (WS)—a rare genetic deficit which results in an unusual cognitive profile of severely impaired spatial representation together with relatively spared language. Studies of spatial representation in this population have shown that even within the broad category of spatial representation, there is uneven sparing and breakdown. The hallmark impairment in people with WS is their performance on visual-spatial construction tasks such as figure copying and block construction (Bellugi,

Marks, Bihrle, & Sabo 1988; Georgopoulos, Georgopoulos, Kurz, & Landau 2004; Hoffman, Landau, & Pagani 2003; Mervis, Morris, Bertrand, & Robinson 1999). To illustrate the severity of this deficit, Plate 2 shows a typical set of copies by two 11-year-olds with WS, in comparison to those of a normally developing 6-year-old child. Clearly, there is a failure among the WS children to recreate even relatively simple spatial relationships. On the other hand, people with WS have preserved capacity to represent the spatial structure of objects, even when the stimuli are briefly presented for identification without surface cues such as color and texture (Landau, Hoffman, & Kurz 2006). They also show strong preservation of face representation, along with a normal, classic 'inversion effect', suggesting that they likely process faces holistically (Tager-Flusberg, Plesa-Skwerer, & Faja 2003). Perception of biological motion and motion coherence are also preserved (Jordan, Reiss, Hoffman, & Landau 2002; Reiss, Hoffman, & Landau 2005), as is spatial language (Lakusta and Landau 2005; Landau and Zukowski 2003).

The puzzling and uneven pattern of spatial breakdown in this syndrome raises significant questions about the nature of normal spatial representation, its developmental profile in normal children, and the nature of breakdown

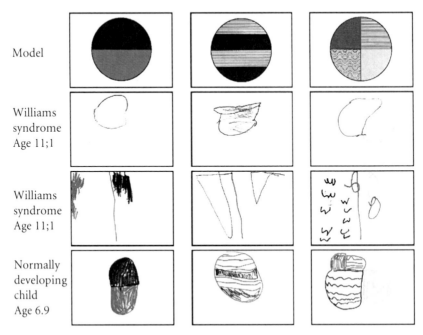

2. Copies of models (row 1) made by children with Williams syndrome (rows 2 and 3) and by one mental age-matched normally developing child (row 4). The models remain visible while the child is copying them. For improved image quality and colour representation see Plate 2.

under genetic deficit. The hallmark pattern of breakdown—copying, block construction, and other visual-spatial construction tasks—raises the possibility that some quite fundamental mechanisms of spatial representation are severely impaired. We will use this chapter to explore whether a breakdown in mechanisms of bodily anchoring (in various forms) might account—either partly or fully—for the pattern of spatial deficit. If people with WS cannot use anchoring to connect their mental representations to the world, this could lead to a variety of serious breakdowns in spatial cognitive functions. By considering this possibility, we will also explore the extent to which such mechanisms can in principle account for the nature of our spatial capacities and, in the case of WS, patterns of spatial breakdown.

To do this, we will tackle a group of phenomena which—at least on the face of it—involve some form of body-world interaction that results in anchoring to the physical world. These phenomena come from very different problem domains, including multiple object tracking, spatial problem solving in block construction tasks, and the use of language to label object parts. Each of these has been argued to involve a kind of physical-spatial 'anchoring' (or direct reference to the world), albeit using different terminology across different problem domains. For example, multiple object tracking tasks have been thought to be accomplished via 'visual indexes', which enable the perceiver to simultaneously mark as many as four objects in an array, and thereby to track these objects as they move through space (Pylyshyn 2000). Block construction tasks have been offered as examples of the visual system's use of 'deictic pointers' to mark spatial locations and thereby allow the perceiver to revisit these locations while solving spatial problems. This capacity might allow people to solve complex problems with a minimum reliance on internal spatial memory (Ballard et al. 1997). And spatial language has long been argued to incorporate spatial biases that follow from the fact that we spend much of our time upright in a three-dimensional world (Clark 1973; Shepard & Hurwitz 1984). Recent proponents in the embodiment framework have even argued for strong effects in the opposite direction, with our spatial actions affected by the particular language we learn (Levinson 1994).

Consistent with arguments for the pervasive importance of embodiment, our selection of tasks and domains spans a broad class of spatial problems. Further, each problem naturally invites the knower to use some kind of anchoring to the physical environment—whether by attentional mechanisms, eye fixations, or posture. However, we will argue that these anchoring mechanisms have significant limits in explaining the spatial breakdown observed in Williams syndrome. Specifically, we will argue that the spatial impairment is not caused by deficits in the mechanisms that help anchor actions and thoughts to the physical world. Rather, we will propose that these anchoring mechanisms are intact in people with Williams syndrome. This indicates that the culprit for breakdown lies in higher-level operations that require abstract internal spatial representations.

We now present a preview of our tasks and findings, which will include successes and failures by Williams syndrome individuals over three domains. In section 7.2, we will explore the capacity to simultaneously track multiple objects as they move through space. We will report success. Specifically, we argue that the basic function of assigning and deploying multiple visual indexes to objects (moving or static) is intact. Thus impairment in this function cannot, by itself, explain the well-documented spatial impairment of WS people. At the same time, we will report a kind of failure: that WS individuals may have a smaller capacity for indexes than normal individuals. This raises the possibility that the indexing function may interact with general attentional mechanisms that might be impaired in WS. In section 7.3, we explore the hallmark impairment of WS by examining performance in block construction puzzles. Ballard et al. (1997) argued that normal adults solve these puzzles by establishing deictic pointers to individual blocks in the puzzle, allowing them to solve complex puzzles without relying heavily on visual-spatial memory. Again, we will report success. Specifically, we find that WS children systematically revisit the relevant portions of the puzzle space via eye fixations as they complete the puzzles. At the same time, we find failure, with severe impairment in their final construction of the block copies. Their failure leads us to examine other possible sources of the problem, and the answer helps shed light on the limits of deictic pointers. Finally, in section 7.4, we explore WS children's ability to mark an object's spatial structure with the terms 'top', 'bottom', 'front', 'back', and 'side'. Once more, we will report success, with WS children showing striking accuracy in establishing markers for an object's six sides, forming a coherent group of spatial part terms for both familiar and novel objects. However, this success is evident only when the structure of the markers is consistent with gravitational upright—that is, 'top' is the uppermost part, in the gravitational or environmental frame of reference. The failures that occur when this consistency is violated give us further insight into the limits of embodied representations of space.

## 7.2  Visual indexes: success and failure in the multiple object tracking task

The idea of visual indexes was first proposed by Pylyshyn (1989) in an attempt to characterize a basic visual function that enters computationally into many high-level tasks of visual cognition. The basic idea is that, prior to carrying out these tasks, the visual system must be capable of 'pointing to' objects in the world. This pointing function is thought to be carried out by a specialized visual mechanism that marks where a particular objects (or 'proto-object') is, and can subsequently refer to these throughout later stages of visual computation (Pylyshyn 2000; Pylyshyn & Storm 1988). The fundamental nature of such a marking function has also been noted by other vision scientists; for example, Ullman (1984) suggests that 'marking' is one of the basic mechanisms necessary for 'visual routines' such as curve tracing.

The original proposed indexing mechanism was called FINST (Fingers of Instantiation; Pylyshyn 1989), drawing an analogy with our real fingers, which can point to a limited number of things simultaneously without representing any information other than 'it's there'. Subsequent theoretical and empirical developments by Pylyshyn and colleagues have suggested several important properties of this mechanism (Pylyshyn 2000; Pylyshyn & Storm 1988; Scholl & Pylyshyn 1999; see also Leslie, Xu, Tremoulet, & Scholl 1998). For example, the mechanism can point to around four indexes simultaneously (but see Alvarez & Franconeri 2007 for evidence that this is related to task demands). Evidence consistent with this idea shows that adults can accurately track up to about four moving objects simultaneously. Second, the mechanism permits people to track a set of moving items as long as the items transform in ways that obey the constraints of physical objects. First, people can track up to four stimuli as they move behind occluders, but cannot do so if the objects shrink and expand at the boundaries of the occluders in ways not characteristic of real physical objects (e.g. by implosion: Scholl & Pylyshyn 1999). Third, people have great difficulty tracking motions of stimuli that follow paths appropriate for substances, thereby violating properties of physical coherence (vanMarle & Scholl 2003).

Perhaps most importantly for us, Pylyshyn (2000) suggests that visual indexing may be a necessary component of many higher-order visual tasks. For example, he proposes that indexing (which allows up to four pointers at a time) may support our ability to subitize, i.e. rapidly 'count' up to four items accurately and with very little decrease in speed with increasing number. He also proposes that indexing may support our ability to carry out visual-spatial copying tasks, such as block construction or drawing. This idea is similar to that proposed by Ballard et al. (1997; see section 7.3). People typically carry out these construction tasks in a piecemeal fashion, by selecting parts of the model puzzle, copying them, then returning to the model to select another part, and so forth. This process requires that a person mark each part of the model as he selects it, in order both to check the accuracy of their copy and to prevent reselecting the same part on the next cycle. The marking could be done by physical marks (such as with a pencil); but Pylyshyn proposes that the visual indexing mechanism generally performs this function quite automatically and naturally.

Given the important spatial role of indexing, it seems quite possible that the visual indexing mechanism is damaged in people with Williams syndrome. If so, it might lead to difficulty in block construction tasks, problems in numerical computation, and deficits in drawing and copying. Both severely impaired block construction performance and drawing deficits have been amply demonstrated in this population (see earlier discussion). In addition, there is speculation that number knowledge is severely impaired, although the exact profile is not yet well understood (see e.g. Ansari & Karmiloff-Smith 2002; O'Hearn, Landau, & Hoffman 2005a; O'Hearn & Landau 2007; Paterson, Brown, Gsödl, Johnson, & Karmiloff-Smith 1999; Udwin, Davies, & Howlin 1996).

In our studies (O'Hearn, Landau, & Hoffman 2005b), we asked whether children and adults with Williams syndrome could carry out a multiple object tracking task (MOT), which Pylyshyn and Storm originally designed as a marker task for visual indexing (see also Scholl & Pylyshyn 1999). Because people with WS have difficulties with a range of visual-spatial tasks, we contrasted performance in the MOT with a Static task, which tested memory for the locations of multiple static objects under testing conditions parallel to those of the MOT. Carrying out both tasks allowed us to ask whether the indexing mechanism exists at all in WS, whether it functions as efficiently as in normally developing children and normal adults, and whether it can be separated from representation of static spatial location. Difficulties on both tasks would suggest impairment in representing and remembering location, whether static or under motion. Difficulties on only the MOT task would suggest breakdown confined only to tracking moving objects. Because this kind of tracking is a marker task for the indexing mechanism, this pattern of performance would suggest that there is breakdown in indexing, but not representation and memory for static location. Alternatively, it could indicate that the indexing function is intact when it is applied to static objects (as in the block construction task: see section 7.3), but shows breakdown when it is applied to moving objects. Further, if there is breakdown in either condition, then detailed patterns of breakdown should reveal whether the mechanism for tracking static or moving objects is completely absent or, if not, how it is different from normal.

Our WS participants included 15 children and adults with a mean age of 18 years. Because WS individuals are typically moderately mentally retarded (mean IQ = 65; Mervis et al. 1999), we compared their performance to the same number of normally developing children who were individually matched to the WS subjects on their mental age (MA matches). This was measured by the children's raw scores on a widely used intelligence test (KBIT: Kaufman & Kaufman 1990). These mental age matched children were, of course, chronologically much younger than the WS subjects (mean age of MA matches = 5 years, 11 months).

Both tasks were carried out on a computer. People first saw eight solid red 'cards', each 1 1/8 in. square, which appeared in non-overlapping but random locations on the screen (see Figure 7.1). These cards then flipped over, revealing cat pictures on one to four of the cards (i.e. targets). People were told to remember which cards had cats, and were given time to study the display, often counting the cats. When people indicated that they were ready, the cards flipped back over, showing their identical solid red sides. At this point, the Static and MOT tasks differed. In the Static task, the cards remained stationary for 6 seconds; in the MOT task, they moved along randomly generated trajectories for 6 seconds. After this period, in both tasks, people pointed to those solid red cards that they thought concealed the target cats. As they did so, the selected card would 'flip over', revealing either a white side (non-target) or a cat (target), together with an auditory meow. The two tasks were counterbalanced, and each was preceded by two practice trials. Each task had 24 trials, evenly and randomly divided among one, two, three, or four target trials.

The results for the Static task showed no differences between the children and adults with WS and their mental age-matched (MA) controls. However, the results of the MOT task revealed that the WS people performed much more poorly than the control children (see Figure 7.2). In particular, the WS group performed worse than the mental age matches on the 3 and 4 target trials in the MOT task. This pattern suggests that the indexing mechanism—which putatively allows tracking of multiple objects as they move through space—may be impaired in people with WS. It also suggests that representing the specific locations of up to four *static* objects is not impaired in people with WS, at least on this task and in comparison to MA children. In follow-up studies, we asked whether the pattern of better performance in the Static task than in the MOT was characteristic of normally developing children who were younger than our MA controls. Normal 4-year-olds did not show the WS pattern. While the 4-year-olds performed similarly to people with WS in the MOT task, the WS group performed *better* than the 4-year-olds in the Static task.

Overall, these findings suggest several conclusions about visual indexing in people with WS. First, these people were impaired at tracking multiple moving objects but not at remembering the static locations of multiple objects. This hints that the indexing mechanism—proposed to support our tracking of

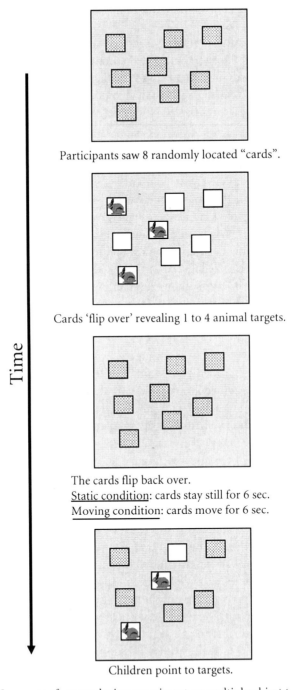

Participants saw 8 randomly located "cards".

Cards 'flip over' revealing 1 to 4 animal targets.

The cards flip back over.
Static condition: cards stay still for 6 sec.
Moving condition: cards move for 6 sec.

Children point to targets.

FIGURE 7.1. Sequence of events during experiment on multiple object tracking, in the static condition and the object tracking condition (adapted from O'Hearn, Landau, & Hoffman 2005b). See text for discussion.

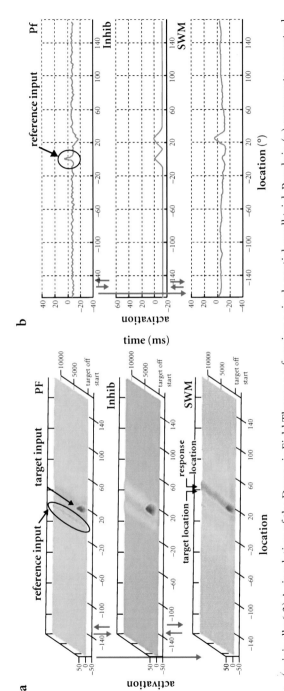

1. (originally 6.2) A simulation of the Dynamic Field Theory performing a single spatial recall trial. Panels in (a) represent: perceptual field [PF]; inhibitory field [Inhib]; working memory field [SWM]. Arrows represent interaction between fields. Green arrows represent excitatory connections and red arrows represent inhibitory connections. In each field, location is represented along the x-axis (with midline at location 0), activation along the y-axis, and time along the z-axis. The trial begins at the front of the figure and moves toward the back. (b) Time slices through PF, Inhib, and SWM at the end of the delay for the model shown in (a). See text for additional details.

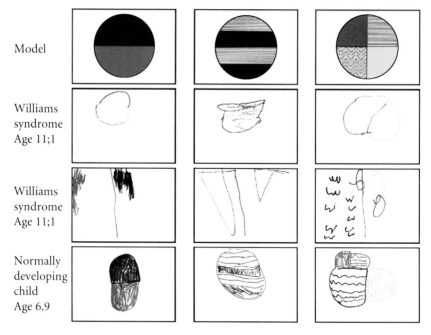

2. (originally 7.1) Copies of models (row 1) made by children with Williams syndrome (rows 2 and 3) and by one mental age-matched normally developing child (row 4). The models remain visible while the child is copying them.

## Manipulate

## Anchor

3. (originally 7.7) In the Manipulate condition, children remove the target objects one at atime from a bag, and proceed to label the object parts as queried (e.g. top, bottom, front, back, side). Children tend to manipulate the objects as they label the parts, thus changing the relationship between the parts, their body, and the environment as they move through each trial. In the Anchor condition, the objects are held in one position in front of the child during the entire trial. The parts remain in stable locations relative to the child and the environment.

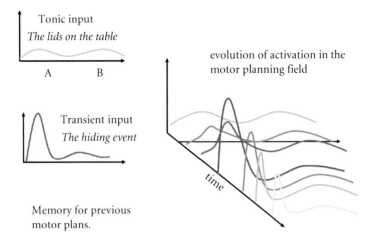

4. (originally 9.2) An overview of the dynamic field model of the A not B error. Activation in the motor planning field is driven by the tonic input of the hiding locations, the transient hiding event, and the memories of prior reaches. This figure shows a sustained activation to a hiding event on the left side despite recent memories of reaching to the right, that is a nonperseverative response.

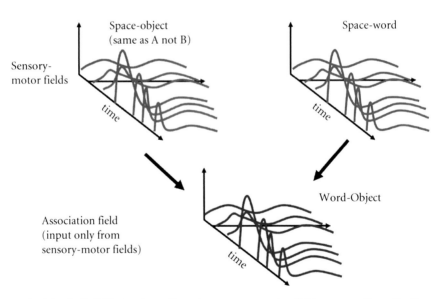

5. (originally 9.4) Illustration of how two sensory-motor fields representing attention and planned action to objects in space and to sounds in space may be coupled and feed into an association field that maps words to objects without represesenting the spatial links of those words and objects.

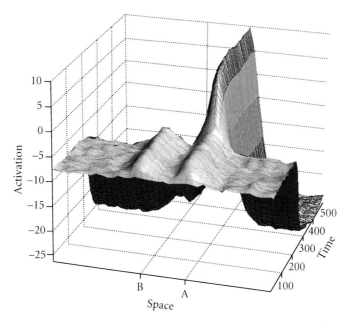

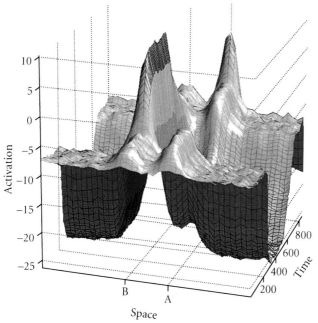

6. (originally 9.3) (A) The time evolution of activation in the planning field on the first A trial. The activation rises as the object is hidden and due to self-organizing properties in the field is sustained during the delay. (B) The time evolution of activation in the planning field on the first B trial. There is heightened activation at A prior to the hiding event due to memory for prior reaches. As the object is hidden at B, activation rises at B, but as this transient event ends, due the memory properties of the field, activation.

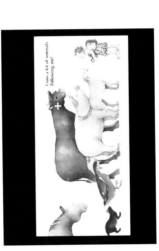

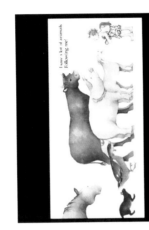

7. (originally 10.2) The snapshots when the speaker uttered "the cow is looking at the little boy" in Mandarin. Left: no non-speech information in audio-only condition. center: a snapshot from the fixed camera. Right: a snapshot from a head-mounted camera with the current gaze position (the white cross).

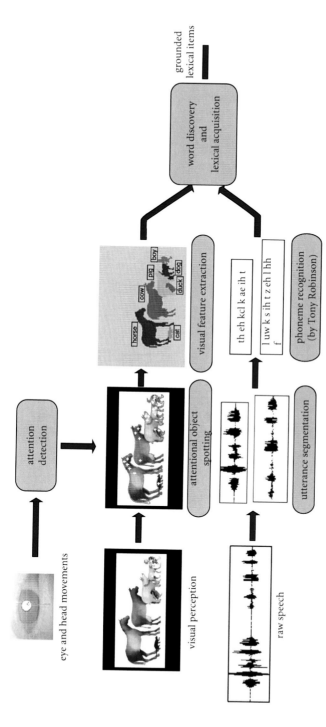

8. (originally 10.6) The overview of the system. The system first estimates subjects' focus of attention, then utilizes spatial-temporal correlations of multisensory input at attentional points in time to associate spoken words with their perceptually grounded meanings.

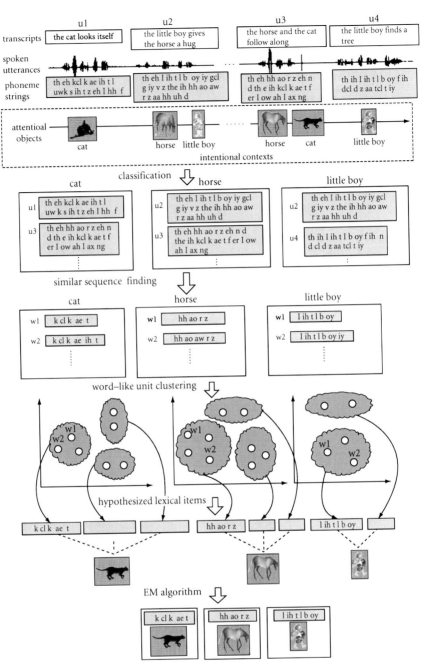

9. (originally 10.7) Overview of the method. Spoken utterances are categorized into several bins that correspond to temporally co-occurring attentional objects. Then we compare any pair of spoken utterances in each bin to find the similar subsequences that are treated as word-like units. Next, those word-like units in each bin are clustered based on the similarities of their phoneme strings. The EM-algorithm is applied to find lexical items from hypothesized word-meaning pairs.

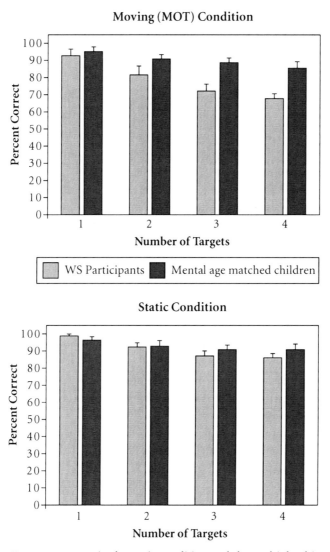

FIGURE 7.2. Percentage error in the static condition and the multiple object tracking condition (adapted from O'Hearn et al. 2005b)

moving objects—may be damaged in some way. At the same time, our WS group was capable of tracking one or two objects at the level of MA children, and was able to track even three or four moving objects at better than chance levels. This suggests that the indexing mechanism is not completely *absent*, but rather, seems to suffer from greater inaccuracy with larger numbers than that of MA children.

What could be the source of this greater inaccuracy? Recently, we have examined the hypothesis that the indexes of WS individuals are more 'slippery' than those of MA children. If an index is present, but then slips off a target, it would probably end up on a spatial neighbor, that is, another object that passes close to it as it moves along its path. To measure the idea of slippage, we identified where the false alarms (i.e. objects incorrectly identified as targets) occurred and computed the distance to the real target. If the false alarms reflect an index 'slipping' off a target, then the distance between the real target and a false alarm should be smaller than the distance between the real target and correct rejections (see O'Hearn, Landau, & Hoffman 2005b).

We discovered that all groups of subjects showed some degree of slippage; that is, at some point during their respective trajectories, their false alarms had passed closer to the real targets than had the correct rejections. However, the distances between false alarms and targets among WS individuals were, overall, larger than those of normal, MA children. This suggests that WS individuals may experience slippage some of the time, but on other occasions may simply be missing an index, which would force them to guess. The idea of having fewer indexes would be consistent with other related evidence from visuo-spatial search tasks (Hoffman and Landau, in preparation). The idea that existing indexes might more easily be 'lost' during the trajectory would be consistent with the idea that WS individuals have an overall impairment in their attentional system (see Brown, Johnson, Paterson, Gilmore, Longhi, & Karmiloff-Smith 2003, for related evidence in WS infants). We are currently examining both hypotheses.

Returning to our broader goal in this chapter, we believe that the evidence from both the Static and the MOT task suggests that the visual indexing mechanism is present in people with Williams syndrome. They can retain the locations of up to four static objects in displays of eight, and they can do so over a retention period of 6 seconds at levels similar to those of normal children who are matched for mental age. They can also track the locations of multiple objects over motion, although they perform reliably worse than their mental age matches for target sets of three or four. Considered in the framework of indexing (Pylyshyn 2000), this evidence indicates that there may be fewer indexes than normal, yielding more random guesses. Still, while people with WS appear limited in their ability to deploy indexes, the fundamental ability to refer to and locate objects in the external world appears intact.

We tentatively conclude that complete failure of the visual indexing mechanism cannot alone explain the documented severe spatial deficit in Williams syndrome. To explore this point more directly, we now turn to the block construction task.

## 7.3  Deictic pointers: success and failure in the block construction task

The idea of 'deictic pointers' was first laid out in a paper by Ballard and colleagues (Ballard et al. 1997), where they proposed that people use the world as an aid to memory, thereby decreasing the burden on internal visual-spatial representations. Deictic pointers were proposed as the mechanism whereby parts of the visual-spatial world could be mentally marked for later use, for example, either to return to the same location or to visit new locations (by avoiding old ones). Ballard et al. documented the importance of this mechanism by studying people's performance in block construction tasks quite similar to those that reveal the hallmark spatial deficit among people with Williams syndrome. These tasks, which are also a common part of standardized IQ tests, require that people replicate multi-block models in an adjacent copy space by selecting candidate blocks from a set presented in a third area (see Figure 7.3).

Intuitively, one might assume that people solve these puzzles by consulting the model just once, then selecting the correct blocks and assembling a replica in the copy area. However, Ballard et al. found that people solve these puzzles in quite a different way. Their central analyses concerned the sequences of eye fixations that people produce as they carried out the task. First, they found that people generally fixated the model much more often than our intuitive description predicts. In fact, people fixated the model twice during the placement of *each block* in the copy area. Initially, subjects fixated a single block in the model, apparently to determine its color. This allowed them to pick up the corresponding block in the parts area. They then revisited the model block to encode the target block's location before finally placing the block in the copy area. Ballard et al. called this approach the 'minimal memory strategy' because subjects apparently use frequent looks to the model instead of trying to commit to memory sets of blocks or the entire model. Ballard et al. argued that this strategy is optional. They found that when the model and copy areas were moved further apart, subjects made greater use of memory, fixating the model less often. They suggested that, when eye movements become 'expensive', as in the case of a model and copy area that are far apart, people resort to storing larger chunks (i.e. multiple blocks) in memory.

The minimal memory strategy depends on maintaining deictic pointers to specific locations in the model. For example, when observers make an initial fixation on a model block, they not only encode its color but also create a pointer to the block. This pointer allows the observer to return to the block at a later time, but does not contain explicit information about the block's color or location. After encoding the color, subjects can retrieve a block from the parts

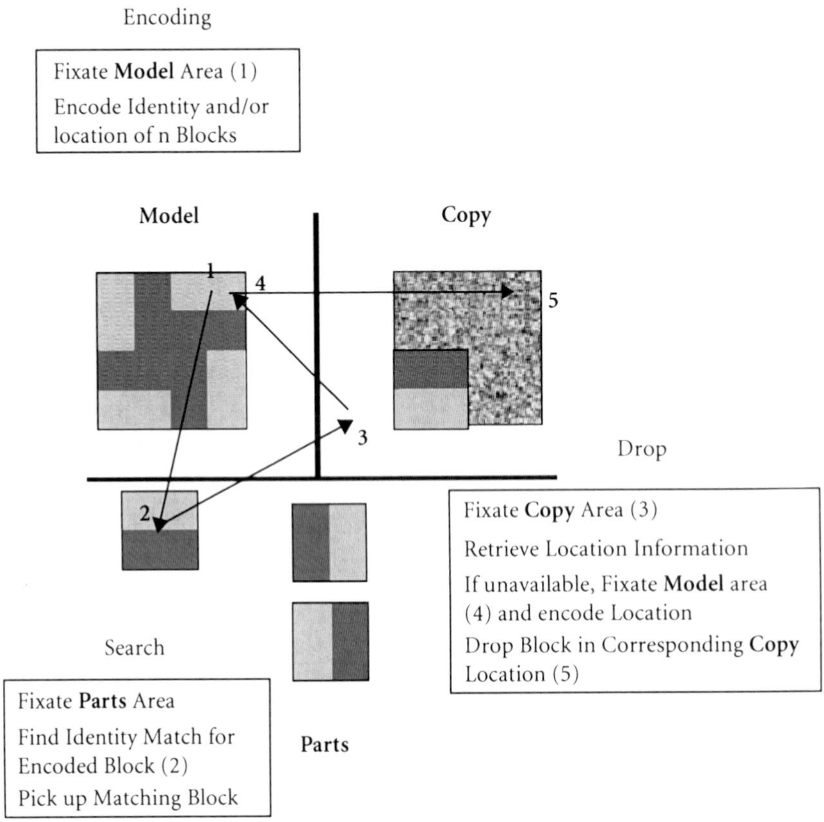

Encoding

Fixate **Model** Area (1)

Encode Identity and/or location of n Blocks

**Model**    **Copy**

5

**Drop**

3

Fixate **Copy** Area (3)

Retrieve Location Information

If unavailable, Fixate **Model** area (4) and encode Location

Drop Block in Corresponding **Copy** Location (5)

Search

Fixate **Parts** Area

Find Identity Match for Encoded Block (2)

Pick up Matching Block

**Parts**

FIGURE 7.3. Sequence of actions required to solve the block construction puzzle (adapted from Hoffman, Landau, & Pagani 2003). See text for discussion.

area; but then they must check that the selected block was in fact the right color. They do this by accessing the pointer to refixate the relevant block in the model. During this second fixation, they can determine the location of the block with respect to the model, allowing them to correctly place the part in the copy area. They can then follow their pointer back to the model once again to choose a neighboring block with which to start the entire cycle again.

Ballard et al.'s proposal invites two possible explanations of the severe difficulties that people with Williams syndrome experience in carrying out block construction tasks. First, it is possible that WS people do not employ deictic pointers as they carry out block tasks. If so, this would prevent the subject from revisiting individual blocks in the model in order to check, first, that a block of the correct color has been chosen from the parts area, and second, that it has been placed in the right location in the copy area. A second related possibility is that, without deictic pointers, subjects would be incapable of using the 'minimal memory strategy', and would be forced to rely on maintenance of spatial information in working memory. Such a strategy of relying on spatial representations in working memory would be a potent source of errors for people with WS, given their well-documented weakness in a variety of working memory tasks (Jarrold, Baddeley, & Hewes 1999; O'Hearn Donny, Landau, Courtney, & Hoffman 2004; Paul, Stiles, Passarotti, Bavar, & Bellugi 2002).

We evaluated these possibilities by examining eye fixations in children with WS and a group of MA controls during a computerized version of the block construction task. We used both solid colored blocks like those in the Ballard et al. study as well as blocks containing visual-spatial structure. For example, blocks could have a horizontal structure, with red on the top and green on the bottom, as well as vertical and diagonal structure (see Figure 7.3 for example with light and dark gray representing red and green). In measures of overall performance (i.e. puzzle solution), WS children were quite accurate on puzzles with solid colored pieces, and close to ceiling. However, they showed characteristic breakdown in performance when puzzles contained pieces with internal structure; as in all studies of block construction, they were reliably worse than the normally developing children who were matched for mental age. In order to test the possibility that children with WS were not using deictic pointers as Ballard et al. described, we examined the children's eye fixations throughout their solution of the puzzles. We found that, for puzzles containing four or fewer pieces, WS and MA normal children used the same strategy identified by Ballard et al., making multiple fixations on the model prior to each placement of a block in the copy area. Despite these similar fixation patterns, however, WS children made many more errors than the MA controls, hinting that something *other* than an absence of deictic pointers was responsible for the breakdown. In larger puzzles (nine pieces), we found a change in strategy among the WS children. They now fixated the model once and then proceeded to place multiple pieces without additional fixations on the model, leading to predictably poor accuracy. This change in strategy, however, appeared to be a *response* to poor performance rather than a *cause*. For example, we found that even on trials in which WS subjects did fixate the model, their performance was barely above chance.

These results suggest that the breakdown seen among WS children in the block task is not due to failures to deploy pointers but rather to failures in constructing and maintaining representations of the block structure and/ or location. We investigated this possibility in two follow-up tasks in which we eliminated the need for deictic pointers altogether, while also testing for the presence of impaired spatial representations. In the Matching task, we

**Matching Task**

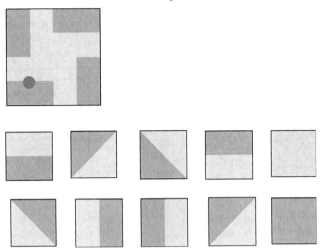

FIGURE 7.4A. Panel seen during the Block Matching task (adapted from Hoffman et al. 2003). People are shown the block puzzle (top panel), with one block marked as target with a large dot. They then must select, from the choices below, the block that matches the target.

**Location Task**

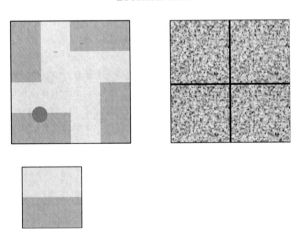

FIGURE 7.4B. Panel seen during the Location Matching task (adapted from Hoffman et al. 2003). People are shown the block puzzle (top panel), with one block marked as target with a large dot. They are also given the correct match in the panel below, and must move this target block into the correct space in the blank model at the right.

designated the relevant block in the model by placing a red disk at its center. Subjects had to choose a matching block from the parts area below the model, just as they had in the full version of the block construction task (see Figure 7.4a). In this new Matching task, however, there was no need to use pointers because there was a salient visual feature that could mediate refixations on the model. We found that WS subjects were severely impaired on this task compared to MA controls, with most of their erroneous choices being mirror reflections of the correct blocks. In a second follow-up task, the Location task, we once again cued a model block with a red disk at its center and placed a single, matching block in the parts area (Figure 7.4b). Subjects were required to move the block into the corresponding location in the copy area, which contained a grid showing possible locations. Once again, WS subjects were impaired relative to controls, despite the salient visual marker on the model obviating the need for pointers. Importantly, the combined performance on the Matching and Location tasks were highly predictive of performance on the full block construction task, suggesting that the key to breakdown was in the representation of spatial structure of blocks and their locations relative to each other.

The results of all three experiments suggest that poor performance by WS participants on the block construction task is not due to a failure to use deictic pointers or to 'slippage' of pointers from their intended locations. Indeed, on virtually all measures of fixation in the full block task, the WS children were comparable to the MA children; and both groups were often comparable to normal adults. The results confirm Ballard's notion that deictic pointers help people to solve complex visual-spatial tasks by 'marking' areas of a space in order to revisit and check periodically. However, in the case of WS, there is severe failure to accurately represent and maintain over time the correct spatial structure of individual blocks and their relative locations within the puzzle, as shown in the Matching and Location tasks. We believe that this shows that the power of deictic pointers can only be realized in the context of an equally powerful system of spatial representation coupled with working memory.

## 7.4 Gravitational anchoring: success and failure with spatial part words

Our final case concerns spatial part words—terms such as 'top', 'bottom', 'front', 'back', and 'side', which encode the parts of objects in an object-centered reference system. These reference systems allow distinct regions of an object to be marked, labeled, and retained over an object's rotation, translation, or reflection in space. Neurophysiological and cognitive studies on animals, normal adults, and brain-damaged patients have shown that the object-centered reference system is used in a variety of non-linguistic spatial tasks (Behrmann 2000; see Landau 2000 for review). Spatial part terms in English (and other languages) appear to capitalize on the existence of these object-centered representations by allowing speakers and hearers to distinctively 'mark' the six 'sides' of an object (Landau 2000; Levinson 1994; van der Zee 1996).

The distinctive linguistic marking of these sides suggests that non-linguistic representations of objects may also have such structure. If so, then theories of visual object representation should probably be enriched to include different sets of axes within an object, and different directions within each axis. Landau and Jackendoff (1993) proposed that the terms reflect a spatial structure that is characterized by three different sets of axes. These include the Generating axis (central or main axis), the Orienting axes (secondary front-back and left-right axes), and the Directing axes (which further distinguish front from back and left from right). If the visual system makes such distinctions, then this could support the acquisition and use of the spatial part terms in English and other languages. Once the terms are applied to an object, a person can hold onto the marked structure as the object moves through space, thus re-identifying, over motions, the same 'top', 'bottom', and so forth.

This description presupposes that geometric structure is the only factor in assigning spatial part terms—an assumption that proves to be questionable (see below). But to the extent that the spatial structure is one important factor in learning and using the terms, this suggests an important role for deployment of indexes or deictic anchoring. Specifically, if the learner must map the set of terms onto a unitary spatial structure, then it seems likely that some kind of visual or mental 'marking' must be deployed. This is necessary to differentially mark different regions of an object as 'top', 'bottom', etc.

Interestingly, using indexes or deictic anchoring to mark each end of the six half-axes of the object could permit spatial terms to be applied to spatial regions of an object and to maintain these marked areas reliably over changes in the object's orientation. Furthermore, an extension of the notion of deictic anchors might lead to quite powerful capacities *if* supplemented with representation of spatial relations that locate each anchor relative to the others. For example, using Pylyshyn's (2000) terms, a group of six markers could constitute a set of visual indexes that would be deployed together to mark the six termini of the three main axes. Once these are marked, then in principle the anchors or indexes might be capable of 'sticking' to the object as it turns or moves. A similar description could be made using Ballard et al.'s 'deictic pointers'.

If the ends are marked in such a way, several things would follow. First, the terms would remain coherently structured over the object's motion: The 'top' and 'bottom' would always be at the opposing end of one axis, whereas each 'side' would be at the opposing end of the secondary axis, and the 'front/back' would oppose on the tertiary axis. This would enable inferences about these terms for novel objects. For example, if one is told where the 'top' of any object is, then the location of the 'bottom' should follow. Second, this pattern of inferences should hold regardless of the position or orientation of the object. For example, if told that a novel object's 'top' is the part lying at the gravitationally lower end of the central axis, one should infer that the object is upside down, with the 'bottom' at the region that is at the gravitationally upper end.

Given these possibilities, the intact ability to use deictic anchors or indexes might enable people with Williams syndrome to learn and use the spatial part terms—at least in limited contexts. An additional ability to carry the *set* of deictic anchors *along with a spatial representation of their locations* would enable people with WS to carry out powerful inferences about the relative locations of spatial part terms, for both familiar and novel objects.

We carried out several experiments to examine these issues (Landau & Kurz, in preparation). We first asked whether people with WS could apply the set of spatial part terms to the correct parts of both novel and familiar objects. We also asked whether the representation of the terms as a spatial whole could support inferences when the objects were in canonical and unusual orientations. A positive answer to the first question would suggest the possible use of deictic anchors to 'mark' the relevant regions. A positive answer to the second question would suggest that these anchors can be *spatially organized* into orthogonal axes and directions within the axes, and maintained as the object appears in unusual orientations.

In a first experiment, we gave children a bag containing a variety of common objects (Figure 7.5). We asked the children to remove the objects one at a time, and while they manipulated the objects, we asked them to indicate the region corresponding to five different spatial part terms: 'top', 'bottom', 'front', 'back', and 'side'. To encourage precision, the children were asked to 'Put your finger on the [spatial part]'. We tested ten children with WS and the same number of normally developing children who were matched to the WS children on mental age. We also tested normal adults, to determine the range of reasonable variation in labeling patterns.

The common objects that we used offer people a number of different coding schemes. For example, cups, cameras, chairs, books, and other common objects have tops and bottoms that are usually established in functional terms. In fact, people's patterns of labeling suggest that they construct 'models' of different objects, with the models following from knowledge of the objects' functions as well as geometry. The parts are labeled in accord with these models.

FIGURE 7.5. Sample objects used in spatial part term experiments (Landau & Kurz 2004). See text for discussion.

Because the functions of different objects can vary so much, application of spatial part terms does not necessarily follow strict geometric principles across different objects—even for normal adults (see Landau & Kurz, in preparation). However, adults do follow spatial constraints on their application of terms. For example, once having decided what part is the 'front', people will then typically use 'back' for the opposing end of the same axis.

In order to examine the separate and joint locations of the five spatial part terms, we coded each response in terms of the region that a subject indicated as he or she was queried with each term. This told us what absolute region of the object was being used for each term, whether the regions overlapped, etc. Then we examined how often opposing terms (such as 'top'/'bottom') were assigned to opposing ends of the same axis. For example, if a person had indicated that the 'top' of a pencil was the eraser tip, then we asked whether he or she assigned the 'bottom' to the graphite end (consistent with an axis-based opposition) or to some other (non-opposing) side, such as the side labeled 'Ticonderoga yellow'.

All groups varied somewhat in exactly what region they designated as the target for a particular term. However, typically, once they assigned a given term to a region, they assigned the opposing term to the region at the opposing end of the axis. Normal adults obeyed this constraint on more than 90% of the trials. Normally developing children also did this on more than 80% of the trials. However, children with WS only did so on about 60% of the trials, indicating that their assignment of one member of a pair did not constrain their assignment of the other member of the pair. This could be due to their failure to appreciate such constraints, or to their forgetting their previous assignments.

In a second analysis, we asked whether, when pairs of terms *were* assigned to the same axis, the pairs 'top'/'bottom' and 'front'/'back' were assigned to axes that were *orthogonal* to each other. For example, once a person had assigned terms 'top'/'bottom' to a single axis, this would constrain their assignment of the other terms to regions of the object: 'front'/'back' would have to be assigned to one of the remaining orthogonal axes, as would side. In accord with our earlier observations about the complex functionality of common objects, even normal adults did not always assign pairs of terms to strictly orthogonal axes. For example, adults might have assigned 'front'/'back' to the same regions of the dollar bill as 'top'/'bottom'—with both 'front' and 'top' mapping onto the region with the picture of George Washington and 'back' and 'bottom' mapping onto the reverse side. Adults, in fact, only assigned the two sets of pairs to orthogonal axes on roughly 60% of the trials. Normally developing children only did so about 40% of the time. But WS children were least likely to assign them to orthogonal axes, doing so on roughly 15% of the trials.

Both of these analyses indicate that the WS children have severe problems maintaining the spatial coherence of the set of terms for these common objects. Why? Using the idea that deictic pointers may be an underlying mechanism of spatial term assignment, there are two possibilities. First, it is possible that WS children have more difficulty assigning these pointers, either one or more, to the parts of an object. But failure to simply *assign* the pointers seems unlikely, since the children found it very easy to follow instructions, putting their finger on different target spatial parts as they were mentioned. And many of the individual assignments were correct in the context of some hypothetical spatial scheme. This indicates that assigning the pointers was probably not the problem. A second possibility is that the pointers did not spatially 'adhere' as a group to the object over time. If this were the case, then the group of markers might not cohere as the child answers sequential queries, resulting in locally 'correct' but globally inconsistent responses.

We tested this possibility by carrying out a different version of the same experiment. This time, half of the objects were tested in the same way as before, having children retrieve objects from the bag one at a time and indicating the different spatial part term regions for each. The other half of the objects, however, were 'anchored' by the experimenter, who held each object stably in front of the child, in a canonical orientation. For example, as shown in Plate 3, a miniature couch was held in the proper orientation for seating an upright doll. If children could assign a spatial scheme to the object via pointers, but not carry this over different orientations of the object, the anchored objects should elicit much more spatially consistent use of the part terms than the non-anchored objects.

## Manipulate

## Anchor

3. In the Manipulate condition, children remove the target objects one at a time from a bag, and proceed to label the object parts as queried (e.g. top, bottom, front, back, side). Children tend to manipulate the objects as they label the parts, thus changing the relationship between the parts, their body, and the environment as they move through each trial. In the Anchor condition, the objects are held in one position in front of the child during the entire trial. The parts remain in stable locations relative to the child and the environment. For improved image quality and colour representation see Plate 3.

Anchoring definitely helped the children with Williams syndrome. When the objects were removed from the bag and manipulated at will, these children still assigned spatial term pairs to opposing ends of an axis only about 60% of the time—roughly the same proportion as we had observed in the first experiment. But when the objects were anchored, these proportions rose to roughly 75%. When we analyzed the children's assignment of term pairs to different orthogonal axes, we found that they did so about 30% of the time in the non-anchored condition, but around 50% of the time in the anchored condition. The normally developing children also showed improvement when the objects were anchored. In the original condition, they assigned terms to opposing ends around 90% of the time and assigned pairs of terms to orthogonal axes around 70% of the time. In the anchored condition, they improved somewhat, with proportions of 95% and 80% respectively. The improvement of the WS children was more dramatic, and suggests that part of their problem in the first experiment may have been the tendency to shift their naming scheme within a single object. Since they freely manipulated the objects as they were asked to indicate different target regions, it is quite possible that the continuously changing orientation of each object made it more difficult for the children to maintain a set of fixed markers for the parts of each object.

In a final experiment, we asked whether the WS children also had difficulty using a coherent set of axes when the objects were anchored, but their orientation was not consistent with gravitational upright. This would allow us to dissociate two possibilities: (1) that physical manipulation of the objects itself was the problem, interfering with the ability to anchor in one orientation, or (2) that mental transformation of the set of terms from some canonical upright was the problem.

We used a set of completely novel objects, all roughly hand size and selected to be non-nameable (at least, to us; see Figure 7.6 for samples). On each trial, the children were told the region for one of the spatial part terms, for example, 'See this part? This is the top.' After having been told this term and shown its region, they were queried on the remaining terms. For example, if they were given 'top', they were then asked to put their finger on the bottom/front/back/side. On half of the trials, the given term was applied to the part that was consistent with the object being held in gravitational upright (see Figure 7.6a). On the other half of the trials, the given term was applied to a part that was not consistent with this orientation (Figure 7.6b). In the first case, children who could apply a coherent spatial scheme to a gravitationally upright object would then be able to assign the remaining part terms. The second case, however, required that they mentally shift the entire set of markers (and corresponding part terms) to a new orientation, not consistent with upright. The question

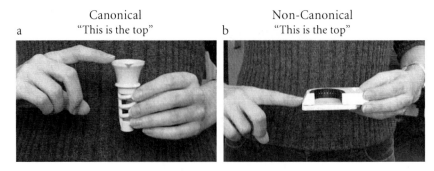

FIGURE 7.6. (a) In the Canonical condition, children were given a single anchor term which was located as if the object were gravitationally upright. (b) In the Non-Canonical condition, they were given the same term, but its location was not a gravitational upright. For example, if the 'top' were at the gravitational side of the object, the object should be understood as rotated horizontally. Children were then asked to indicate the locations of the remaining spatial part terms (e.g. 'Where is the bottom?'). All objects were novel, hence there were no a priori 'correct' locations for the different part terms.

was whether WS children would have special difficulty dealing with the set of terms when they had to assign them to spatial locations inconsistent with the object's canonical orientation.

In the Canonical orientation condition, both WS children and normally developing children performed at ceiling levels, assigning pairs of terms to spatially opposing ends of their axes on about 90% and 100% of the trials, respectively. Both groups also had more difficulty in the Non-canonical condition, but the WS children suffered more. The normally developing children still assigned terms as spatial opposites about 90% of the time, whereas the WS children fell to 70% of the time. When we examined errors, we found that the WS children were much more likely to assign terms to a region that occupied the end of an already 'used' axis. The normal children made few errors, but when they did, they were predominantly assigning the term to its canonical region, e.g. the 'top' to the region at gravitational upright.

As a whole, this evidence points to success and failure in people with WS. They were successful in assigning spatial part terms to familiar objects and to novel objects that were presented in their gravitationally upright orientation (as defined by the location of the 'top'). Failures, however, were seen in the coherence of the relative locations of the spatial terms' regions when the object was no longer in a canonical orientation. This could occur if the object was being manipulated by the child (hence the improvement under conditions of physically anchoring the object). It could also occur if the object's canonical orientation was changed by assignment of the 'top' to a non-typical location on the object. This pattern suggests that deictic anchors may play an important role in allowing assignment of the terms to objects, but that much more is required to enable flexible and unitary coding of spatial parts. What appears to be needed is a coherent spatial representation of the object regions that are distinctively marked by the different terms—as well as the capacity to mentally transform (e.g. rotate) this structure as the object's orientation changes.

## 7.5 Summary and conclusions: the role of embodiment in spatial cognition

Our discussion has focused on the notion that various kinds of embodiment can provide powerful aids to spatial cognitive tasks, but that these aids are not sufficient for the computation or transformation of spatial relationships. The dramatic examples of spatial breakdown in people with WS invite the question of whether failures to use such embodiment aides might play a crucial role: If mechanisms such as indexing, deictic pointing, and anchoring all are crucial to the solution of a broad range of spatial tasks, then deficits in these functions would naturally lead to widespread spatial deficits. Such a pattern of findings would suggest that these embodied mechanisms are both necessary and sufficient for many aspects of spatial cognition.

However, we have presented evidence from three domains that suggests limits to the importance of embodied mechanisms. The spatial deficit in WS does not appear to be accounted for by deficits in indexing, deictic pointing, or anchoring. To the contrary, in each case we have examined, these mechanisms appear to be engaged during marker tasks, and appear to assist people with WS just as they assist normally developing children and normal adults. However, even when these mechanisms are present, we still see severe spatial deficits among people with WS, suggesting that the explanation for these deficits lies elsewhere.

We have proposed that understanding the deficit requires that we think in a different direction—towards the notion that spatial representations are themselves impaired. In the case of the block task, these spatial representations are required for people to correctly identify individual blocks, remember them, and place them in correct locations. In the case of spatial terms, these spatial representations are required for people to use the *collection* of terms coherently, as objects move through space and change their orientation.

In closing, we acknowledge that the notion of spatial representation is highly abstract and, at present, not well understood. However, we submit that there is nothing in the world that 'gives us' spatial representations, nor do these come 'for free', even if we acknowledge that our bodies are anchored in three-dimensional space. Our mental representation of space is a crucial, necessary component of our capacity to carry out even disarmingly simple tasks such as copying block models and indicating the top and bottom of an object. These representations allow us to go far beyond the confines of our body and its physical context, untethering us from the world. Ultimately, it is this untethering that accounts for the power of human spatial knowledge.

## Acknowledgements

We gratefully acknowledge the participating children, adults, and their families. We also thank the Williams Syndrome Association, who helped us locate families of WS participants, and Gitana Chunyo, who assisted in some of the work reported herein. This work was made possible by grants from the March of Dimes Foundation (12-FY0187; FY0446) and the National Science Foundation (BCS 0117744) to B. Landau and the National Institutes of Health (NICHHD F32 HD42346) to K. O'Hearn.

# 8

# Encoding Space in Spatial Language

LAURA A. CARLSON

## 8.1 Introduction

The mapping of language onto space is a topic of interest in many disciplines of cognitive science, including neuroscience (e.g. Farah, Brunn, Wong, Wallace, & Carpenter 1990; Shallice 1996; Stein 1992); cognitive psychology, including psycholinguistics (e.g. Carlson-Radvansky & Irwin 1993; 1994; Clark 1973; Garnham 1989; Landau & Jackendoff 1993; Levelt 1984; 1996), crosslinguistic work (e.g. Brown & Levinson 1993; Casad 1988; Emmorey & Casey 1995; Langacker 1993; 2002; Levinson 1996; 2003; Regier 1996) and attention (e.g. Logan 1995; Regier & Carlson 2001); linguistics (e.g. Jackendoff 1983; 1996; Vandeloise 1991); philosophy (e.g. Eilan, McCarthy, & Brewer 1993); and computer science (e.g. Gapp 1994; 1995; Herskovits 1986; Schirra 1993). One of the reasons that this area has received so much attention is due to the following puzzle. Human beings share a common spatial experience, defined by living in a three-dimensional world, being subject to the forces of gravity, having our perceptual apparatuses and our direction of locomotion oriented in a given direction, and so on (Clark 1973; Fillmore 1971). Nevertheless, there is considerable variability across languages in the way in which we talk about space. To address this puzzle, research has focused on linguistic spatial descriptions as a means of understanding which linguistic properties are associated with which spatial properties.

The examination of linguistic spatial descriptions within this vast body of work can be organized along a continuum, with one end anchored by research focusing on the linguistic properties of the mapping (e.g. Langacker 1993; 2002; Landau & Jackendoff 1993; Talmy 1983) and the other end anchored by research focusing on properties of the spatial representation (e.g. Eilan et al. 1993; Farah et al. 1990). Toward the middle of the continuum is research that examines the interface. Typically, the empirical work at the interface has focused on how language is mapped onto space (e.g. Carlson-Radvansky &

Irwin 1993; Carlson-Radvansky & Jiang 1998; Hayward & Tarr 1995; Logan 1995). For example, consider utterance (1) as a description of a picture containing a fly and an overturned chair.

(1) The fly is above the chair.

Successful mapping requires determining how the features central to the meaning of 'above' such as orientation and direction are assigned within the picture, particularly when conflicting assignments based on the environment versus based on the top side of the chair are possible (e.g. Carlson-Radvansky & Irwin 1993; 1994).

Very little work has taken the opposite approach, that of asking how space is mapped onto language. The goal of the current chaper is to overview three lines of research from my lab that have taken this approach, focusing on which spatial properties are encoded by virtue of processing spatial language. Section 8.2 provides the theoretical framework within which this question is being asked, introducing the concept of a reference frame and its associated parameters. Section 8.3 presents evidence for the encoding of a particular type of spatial information (distance) during the processing of spatial descriptions. Section 8.4 more closely examines the sources of information that serve to define the distance that is encoded. Section 8.5 examines distance as applied within real space. Finally, section 8.6 discusses the implications of these data for the interface between language and space more generally.

## 8.2  At the language and space interface: the use of a reference frame

Imagine the following scenario. You are late for work and are searching among the objects on your kitchen countertops for your keys. Your significant other spots them and provides you with the spatial description in (2).

(2) Your keys are in front of the toaster.

A successful understanding of this seemingly simple utterance depends in large part on your ability to establish a link between the linguistic elements in the utterance and your perceptual representation of the scene at hand. That is, the relevant objects have to be identified, linking 'keys' and 'toaster' with their referents. In addition, their roles have to be correctly inferred, with the toaster (more generally, the *reference object*) serving to define the location of the keys (more generally, the *located object*). Moreover, the spatial term 'front' must be mapped onto the appropriate region of space surrounding the reference object. Finally, the goal of the utterance must be correctly interpreted, with the recognition that the speaker intends to assist your finding the keys by reducing the search space to an area surrounding the reference object whose location is presumably known. This mapping is thought to take place within a representation of the scene, context, and goals, such as a situation model (Tversky 1991; Zwaan 2004), a simulation (Barsalou 1999), or a mesh (Glenberg 1997).

Understanding a spatial term such as 'front' requires interpreting it with respect to a reference system (Shelton and McNamara 2001), a family of representations that map the linguistic spatial term onto space. There are different types of reference system, and these systems serve a variety of different functions in language and cognition (Levinson 1996; 2003). Here we focus on the use of a *reference frame*, a particular type of reference system, and certain classes of spatial terms, including projective spatial terms such as 'left' and 'above' and topological terms such as 'near'. According to Coventry and Garrod (2004), *projective* terms are those that convey information about the direction of a target with respect to another object, whereas *topological* terms convey static relations such as containment ('in') or support ('on'), or proximity ('near'). During apprehension, the reference frame is imposed on the reference object, with the space around the object configured via the setting of a number of *parameters* including orientation, direction, origin, spatial template, and scale (Logan & Sadler 1996). The *orientation* parameter refers to the association of a set of orthogonal axes with the vertical (above/below) and horizontal (front/back and left/right) dimensions. In utterance (1) the competition in defining 'above' on the basis of the environment or on the basis of the top-side of the chair is an example of the different sources of information that can be used to set the orientation parameter. The *direction* parameter specifies the relevant endpoint of a given axis (i.e. the front endpoint versus the back endpoint of the horizontal axis). The *origin* indicates where the reference frame is imposed on the reference object. This could be at the center of the reference object or biased toward a functionally important part (Carlson-Radvansky, Covey, & Lattanzi 1999; Carlson & Kenny 2006). The *spatial template* parses the space around the reference object into regions for which the spatial term offers a good, acceptable, or unacceptable characterization of the located object's placement (Carlson-Radvansky & Logan 1997; Logan & Sadler 1996). The *scale* parameter indicates the units of distance to be applied to space. This parameter has not been extensively studied, and is not clearly defined. For example, labeling the parameter 'scale' presumes a distance that is demarcated in a fixed set of intervals. This has not been tested. Accordingly, in the remainder of this chapter, I will refer to this as the *distance* parameter. The research discussed here more closely examines the distance parameter, exploring both the conditions under which it is set and the sources of information that are used to set it.

## 8.3 Mapping space onto language: the necessity of the distance parameter

### 8.3.1 *Is distance encoded?*

Logan & Sadler (1996) argue that not all spatial terms require all parameters of a reference frame. For example, 'near' may require distance and origin, but not direction and orientation. In support of this, Logan & Sadler (1996) asked participants to draw a located object near a reference object. Placements occurred at a relatively constant (and small) distance in all directions from the reference object, indicating that a specific direction was not implied. Similarly, Logan & Sadler (1996) argue that projective spatial terms such as 'front' or 'left' require direction, orientation, and origin but not scale. In support of this, when asked to draw a located object to the left of a reference object, placements occurred at a relatively constant direction (leftward) at a variety of distances. The assumption underlying these claims about 'near' and 'left' is that the linguistic features of the spatial term dictate which parameters of the reference frame are applicable. Within this view, because projective spatial terms convey direction, they make use of the orientation and direction parameters. However, because they do not explicitly convey distance (in the same manner, for example, as 'near'), then they have no need of the distance parameter. Note that this view is consistent with the approach of mapping language onto space, in that the focus is on how linguistic elements within the term are used to configure space.

If examined through the approach of mapping space onto language, however, this view could be too restrictive. Within this approach, aspects of space may be encoded because they are important more generally for processing the spatial description, for example, because they are consistent with goals or task demands. This is compatible with the view of language as a joint activity between two interlocutors for the purpose of accomplishing a goal (Clark 1996). Returning to the example of the location of the keys in utterance (2), the goal of the description was to assist me in finding the keys, presumably so that I could then pick them up and leave for work. In this case, once I locate the keys, their distance from the toaster becomes available. Encoding such a distance would be relevant to me, because such information would facilitate subsequent action on the keys. Therefore, the setting of the distance parameter in the context of processing the spatial term 'front' would be adaptive, even though 'front' may not itself explicitly convey a distance (but see section 8.4). In support of this idea, in the visual attention literature, Remington & Folk (2001) argue for the selective encoding of information from a perceptual display that is consistent with task goals. Because distance is relevant to locating the object, it would presumably be encoded.

8.3.2  *Empirical evidence for encoding distance*

Carlson & van Deman (2004) examined whether spatial terms such as 'left' or 'above' make use of the distance parameter of a reference frame to encode the distance between the located and reference objects during the processing of spatial language. They used a sentence/picture verification task (Clark & Chase 1972) with sentences containing these spatial terms, such as 'A is above B', and displays containing pairs of letters that were placed a short (about 3° of visual angle) or long (about 8° of visual angle) distance apart. The task of the participant was to determine whether a given sentence was an acceptable description of the spatial relation between the letters. Response times and accuracy associated with making this judgement were recorded. A critical feature of the design was that trials were paired, consisting of primes and probes, and the distance between the letters in the display was either held constant or varied across the prime and probe trials of a given pair. Sample prime and probe displays illustrating these conditions are shown in Figure 8.1.

The underlying logic of the task was that if interpreting the spatial term on the prime trial involved encoding the distance between the located and reference object via the distance parameter of the reference frame, then processing should be facilitated when the same distance setting could be used on the probe trial, relative to when a different setting was required. The main dependent measure was the amount of savings observed on probe trials relative to prime trials, operationally defined as a difference scored by subtracting response time on the probe trial from the response time on the prime trial. When the distance matched between prime and probe trials, we expected savings, expressed as a positive difference. However, when the distance mismatched between prime and probe trials, there should be no such savings. We focused on different scores as the primary measure of interest because response times on any given prime and probe trial are susceptible to the effects of distance on that trial. Indeed, we found that short-distance prime and probe trials were responded to significantly faster than long-distance prime and probe trials. This effect is not informative as to whether distance is maintained; rather, it only shows a difference in processing different distances. To assess whether distance is maintained, one needs to look at the consequences of maintaining a given distance on a given trial for processing on a subsequent trial.

Note that the identity of the letters and their placement within the display was changed across prime and probe trials within a pair. Thus, any facilitation can be attributable to maintaining the distance between prime and probe trials. We also manipulated whether the spatial term matched across prime and probe trials. This allowed us to assess the level at which the distance parameter may be set. Reference frames are hierarchical structures, with the endpoints of an axis

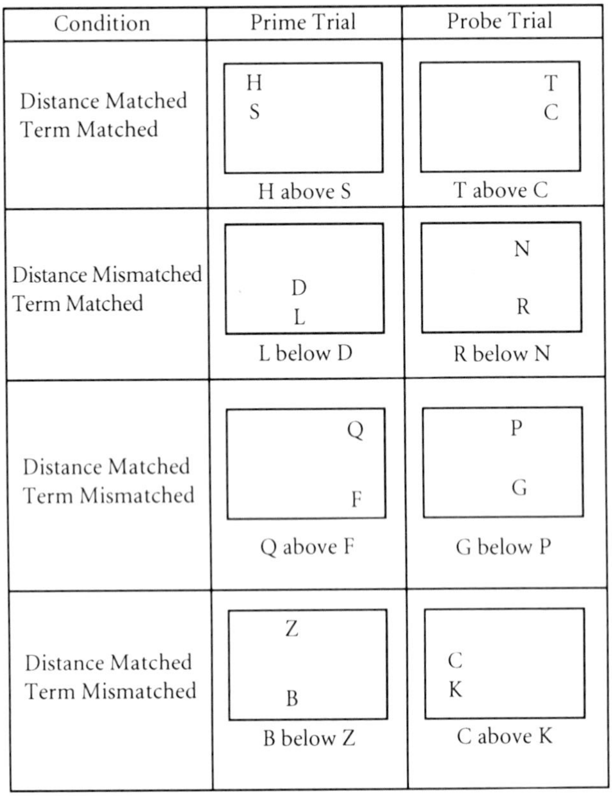

| Condition | Prime Trial | Probe Trial |
|---|---|---|
| Distance Matched Term Matched | H<br>S<br><br>H above S | T<br>C<br><br>T above C |
| Distance Mismatched Term Matched | <br>D<br>L<br>L below D | N<br><br>R<br>R below N |
| Distance Matched Term Mismatched | Q<br><br>F<br>Q above F | P<br><br>G<br>G below P |
| Distance Matched Term Mismatched | Z<br><br>B<br>B below Z | <br>C<br>K<br>C above K |

FIGURE 8.1. Sample displays for the vertical axis and spatial terms 'above' and 'below' that illustrate critical pairs of trials, plotted as a function of distance (matched or mismatched) and term (matched or mismatched) across prime and probe trials. Equivalent displays with the letters horizontally aligned and sentences containing the spatial terms 'left' and 'right' were also used.

(e.g. 'above' and 'below') nested within a particular axis (e.g. vertical) (Logan 1995). Carlson-Radvansky & Jiang (1998) observed that inhibition associated with selecting a particular type of reference frame was applied across the axis of a reference frame, encompassing both endpoints. Accordingly, if the distance parameter is set within an axis so that it applies to both endpoints, then facilitation should be observed when the terms match (e.g. 'above' on the prime trial and 'above' on the probe trial), and when the terms mismatch (i.e. 'above' on the prime trial and 'below' on the probe trial). In contrast, if the distance is set at the level of the endpoint (i.e. tied to a particular spatial term), then we should only observe facilitation when the spatial terms match across prime and probe trials.

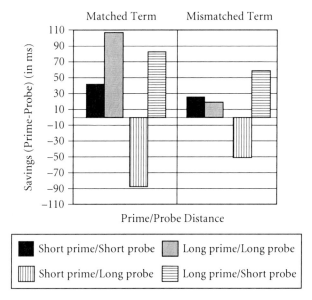

FIGURE 8.2. Savings (prime trial—probe trial) (in msec.) as a function of whether the distance and spatial term matched or mismatched across prime and probe trials in a spatial description verification task. Positive values indicate savings.

Savings are shown in Figure 8.2 as a function of distance (matched versus mismatched) and the spatial term (matched or mismatched) across prime and probe trials. Consider first the conditions in which the distance matched between prime and probe trials (i.e. short prime/short probe and long prime/ long probe). There was significant positive savings, both when the spatial terms matched and when they mismatched. We attribute this facilitation to the maintenance of distance across the prime and probe trials. Note that the size of the savings was smaller when the terms mismatched, indicating a potential additional benefit due to repeating the spatial term across the pair of trials. Now consider the conditions in which the distance mismatches. This pattern of data is accounted for by the fact that on both prime and probe trials, response times on short-distance trials were faster than response times on long-distance trials. Thus, in the short prime/long-distance probe condition, the negative difference is due to subtracting a slower response time associated with the long-distance probe from a faster response time associated with the short-distance prime. The positive difference observed in the long prime/ short probe condition is due to subtracting a faster response time associated with the short-distance probe from a slower response time associated with the long-distance prime. The observation of savings when distance between prime and probe trial matched suggests that distance is encoded and maintained. Moreover, the distance parameter seems to be set within an axis, encompassing both endpoints, as indicated by facilitation when terms mismatched.

An additional experiment was conducted to replicate this effect, and to determine whether the distance effect would operate across axes. Reference frames consist of a set of axes, and it is possible that the distance setting associated with one axis would also be applied to the other axes. To test this, we use spatial terms on prime and probe trials that referred to different axes, for example using 'above' on the prime trial and 'left' on the probe trial. The conditions were similar to those for the mismatched spatial term that are shown in Figure 8.1, except that within a prime/probe pair, one trial used the spatial terms 'above' and 'below' with letters vertically arranged and the other used the spatial terms 'left' and 'right' with the letters horizontally arranged. As in Figure 8.1, distance was either matched or mismatched across the prime and probe trials. If the distance parameter operates at the level of the reference frame (such that a setting associated with the horizontal axis would apply to the vertical axis, or vice versa), then significant savings should be observed when the distance setting matches relative to when it mismatches. The results are shown in Figure 8.3. When the distance matched across prime and probe trials, savings were observed. When the distance mismatched, there was a negative difference when the prime was a short-distance trial and the probe was a long-distance trial, and a positive different when the prime was a long-distance trial and the probe was a short-distance trials. These latter effects are due to short-distance trials in general being responded to faster than long-distance trials.

The claim thus far is that distance is maintained in the context of processing these spatial terms. However, there are alternative explanations. For example, the effect could be due to more general perceptual or attentional processes. On an attentional account, within this task attention must move from one object to the next. As such, when the distance that attention moves is the same across prime and probe trials, there would be facilitation, consistent with the savings that we have observed. On the one hand, this would challenge the idea that the encoding of distance is tied to the processing of the spatial term, because if this more general account is correct, then one would expect to see similar effects in a task with the same attentional components that does not involve spatial language. On the other hand, this is not necessarily a competing hypothesis. Indeed, Logan (1995) has argued that attention is involved in the computation of spatial language (see also Regier & Carlson 2001; for a review of work on attention and spatial language, see Carlson & Logan 2005). Therefore,

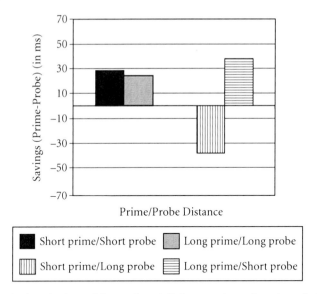

FIGURE 8.3. Savings (prime trial—probe trial) (in msec.) as a function of whether the distance matched or mismatched across prime and probe trials in a spatial description verification task. Note that the term always mismatched across prime and probe trials, because they were drawn from different axes (e.g. above on prime, left on probe; left on prime, below on probe).

attention could serve as the mechanism by which the distance parameter is set. Note that the argument is not that attention is not involved in this task; it certainly is. Rather, the argument is that distance is relevant for the processing of spatial language, and in that context, the distance that attention moves may be encoded.

We conducted two additional experiments to determine whether the distance effect would be observed within tasks that did not involve the processing of spatial terms. In the 'and' experiment participants verified sentences of the form 'A and B', judging whether the letters in the sentence were present in the display. Displays from the previous experiments were used, with letters arranged horizontally or vertically at short or long distances, with the distance matched or mismatched between prime and probe trials. This task shares many of the constituent steps of the spatial language task used previously: namely, both objects need to be identified and verified against the sentence, and attention presumably moves between the objects. However, with respect to task goals, within the 'and' task there is no obvious reason for which distance would be relevant, as the goal is not one of localization, and therefore it is not likely to be encoded. The results are shown in Figure 8.4. When the distance matched between prime and probe trials, no savings was observed. When the distance mismatched, the effects observed were consistent with the previous experiments, and can be explained by slower processing on long-distance trials than short-distance trials.

In the 'bigger'/'smaller' experiment, participants verified sentences of the form 'A is bigger than B' for displays similar to those used in the previous experiments in which letters appeared horizontally or vertically arranged, at a short or long distance, with the distance matched or mismatched across prime and probe trials. In addition, the size of one letter was made slightly bigger or smaller. This task contains even more overlap with the spatial language task, involving identification of the letters, moving attention between them, and making a relational judgement that is spatial in nature but does not involve linguistic terms. As in the 'and' task, it is not clear why distance would be relevant within the size verification task; accordingly, we expected to observe no significant savings when the distance matched across prime and probe trials, in contrast to the findings of the spatial language task. The results are shown in Figure 8.5, broken down as a function of whether the size relation to be judged ('bigger' or 'smaller') matched across prime and probe trials

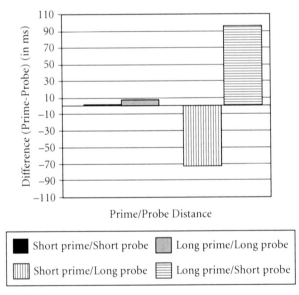

FIGURE 8.4. Savings (prime trial—probe trial) (in msec.) as a function of whether the distance matched or mismatched across prime and probe trials in the 'and' task.

(i.e. a 'bigger' judgement followed by a 'bigger' judgement) or mismatched (i.e. a 'bigger' judgement followed by a 'smaller' judgement). When the relation was the same, there was a savings when the distance matched; however, when the relation was different, there was either no benefit or a negative difference. Accordingly, the effect observed in the matched relation condition appear to be tied to the processing of the particular relation ('bigger' or 'smaller'), and not due to maintaining the distance between prime and probe trials *per se*. This is particularly evident when comparing the pattern of data from this experiment with the comparable conditions in the spatial language experiments in which the spatial terms mismatched between prime and probe, as shown in Figures 8.2 and 8.3. In the case of the spatial language task, savings was observed even when the spatial term was not repeated; in the size judgement task, however, savings were dependent upon repeating the judgement.

In summary, aspects of space such as the distance between two objects are encoded and mapped onto representations used in the processing of spatial language such as the distance parameter of a reference frame. This is consistent with the view that aspects that are relevant to a task goal are selected for encoding (Remington & Folk 2001). However, such information was not retained in

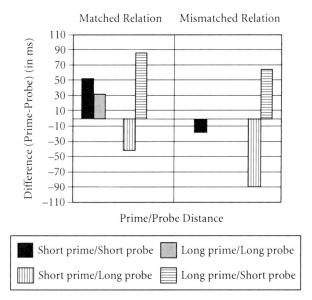

FIGURE 8.5. Savings (prime trial—probe trial) (in msec.) as a function of distance (matched or mismatched) and size relation (matched or mismatched) across prime and probe trials in the size relation task.

tasks that shared many of the constituent processes but did not involve spatial language. This indicates that the effect is tied to spatial language *per se*, and not to a more general cognitive process. If this type of analysis is correct, then the parameters of a reference frame that are deemed relevant for a particular spatial term are not tied to features of the term itself but to the nature of the task in which the term is being used. Thus the fact that projective terms convey direction but not distance is not sufficient for claiming that therefore only direction and not distance information are encoded in the processing of these terms. In an analogous manner, 'near' may explicitly convey distance but not direction and orientation (Logan & Sadler 1996). Nevertheless, consider apprehension of the utterance in (3).

(3) The keys are near the toaster.

It has been shown that such an utterance using 'near' is preferred over one using highly confusable directional terms such as 'left' (Mainwaring, Tversky, Oghishi, & Schiano 2003). Nevertheless, once the keys are located, their orientation and direction vis-à-vis the toaster are available, and can be easily encoded and maintained for future action on the object. Indeed, extending the current methodology, recently we showed that direction information is encoded during the processing of these topological terms (Ashley & Carlson 2007).

## 8.4  Sources of information for setting the distance parameter

### 8.4.1  *The role of object characteristics*

In the experiments from Carlson and van Deman (2004) described in section 8.3, the distance that was encoded was derived from the perceptual displays. This was a necessary feature of the design, in that we needed control in order to match or mismatch the distances across prime and probe trials. However, it is likely that the value assigned to the distance parameter can also be set by other sources of information, such as the characteristics of the objects being related. This idea is supported on both theoretical and empirical grounds. Theoretically, Miller & Johnson-Laird (1976) have argued that the representation of the location of an object contains an area of space immediately surrounding the object, referred to as its penumbra or region of interaction (Morrow & Clark 1988; see also Langacker 1993; 2002). Two objects are said to be in a spatial relation with each other when their regions of interaction overlap. The size of these regions is said to vary as a function of object characteristics. Indeed, Miller & Johnson-Laird argue that objects evoke distance norms that represent typical values associated with their interactions with other objects. Empirically, such object characteristics have been found to influence the setting of other parameters of a reference frame, including the origin and spatial templates (Carlson-Radvansky et al. 1999; Franklin, Henkel, & Zangas 1995; Hayward & Tarr 1995) and orientation and direction (Carlson-Radvansky & Tang 2000; Carlson-Radvansky & Irwin 1993; 1994; Coventry, Prat-Sala, & Richards 2001). In addition, Morrow & Clark (1988) observed systematic object effects on the denotation of the verb *approach*. Participants in Morrow & Clark (1988) were given sentences as in (4).

(4) The squirrel is approaching the flower.

The task was to estimate how far the squirrel (the located object) was from the flower (the reference object). The important finding was that distance estimates varied as a function of the size of the objects, with larger objects being estimated as being farther apart than smaller objects. This effect is important because it suggests that, even for terms that may convey distance explicitly as part of their definition, the value that is conveyed is vague. It is more likely that a range of values is implied, with the actual value selected from this range on the basis of object characteristics.

Carlson & Covey (2005) asked whether the same type of object effects would be observed with spatial terms, both those that seemed to explicitly convey a distance, like 'near' and 'far', and those that did not explicitly convey a distance, like 'left' and 'front'. Given the evidence that distance is relevant to the processing of spatial terms (Carlson & van Deman 2004), it seemed likely that the value that was encoded would be dependent upon characteristics of the objects, regardless of whether the term itself conveyed distance. To address this, we used the paradigm developed by Morrow & Clark (1988) in which participants were provided with pairs of sentences. The first sentence provided a setting, and described a perspective onto a scene, as in (5).

(5) I am standing in my living room looking across the snow-covered lawn at my neighbor's house.

The second sentence spatially related two objects occurring within the scene. Different versions of the sentence were used to systematically manipulate the size of the located and reference objects, as in sentences (6–9).

(6) The neighbor has parked a snowblower in front of his mailbox.
(7) The neighbor has parked a snowblower in front of his house.
(8) The neighbor has parked a snowplow in front of his mailbox.
(9) The neighbor has parked a snowplow in front of his house.

Sentence (6) uses a small located object and a small reference object; sentence (7) uses a small located object and a large reference object; sentence (8) uses a large located object and a small reference object; sentence (9) uses a large located object and a large reference object. Each participant saw only one version of each sentence. The task was to estimate the distance between the objects in feet. Note that there was no visual presentation of a scene containing these objects; thus, the scene had to be imagined, and the distance computed on the basis of this conceptual representation. This design makes it likely that participants would make use of norms associated with the particular objects as a way of computing this distance (Miller & Johnson-Laird 1976). The interesting question is whether these distance norms would change as a function of the objects and their sizes. Different sets of participants provided estimates for different pairs of spatial terms, including 'front'/'back', 'near'/'far', 'left'/'right'. and 'beside'/'next to'.

Mean distance estimates as a function of the size of the located and reference objects are shown in Figure 8.6. The critical finding was that distance estimates associated with smaller objects were significantly smaller than estimates associated with larger objects, with this pattern observed for both located and reference objects. Moreover, the effect seems to be additive, with the smallest estimates associated with small located objects in combination with small reference objects, the largest estimates associated with large located objects in

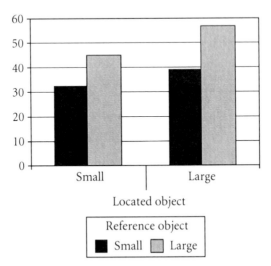

FIGURE 8.6. Distance estimates as a function of reference object size and located object size, collapsing across spatial term.

combination with large reference objects, and the other two conditions falling in between these extremes. Note also that the reference object effect seems to be stronger than the located object effect. This would suggest that the priority is given to the distance norms that define the region of interaction around the reference object. This makes sense in that the goal of a spatial description is to narrow down the search region for a target object to a region immediately surrounding a reference object, with the expectation that the located object falls within the reference object's penumbra or region of interaction (Langacker 1993; 2002; Miller & Johnson-Laird 1976; Morrow & Clark 1988).

### 8.4.2  *The role of the spatial term*

The results thus far suggest that the distance parameter of a reference frame can be set by features independent of the spatial term, including information from a perceptually present display and characteristics of the objects being related. However, this does not necessarily rule out a contribution of the spatial term itself. In the case of topological terms that explicitly convey a range of distances, this most certainly would be the case. For example, 'near' reduces the range of distances from the entire scene to those in close proximity to the reference object, with other factors translating 'close proximity' into an actual value. It is also possible that spatial terms that do not explicitly convey a distance may nevertheless contribute to the setting of the distance parameter, by virtue of the distances that are typically invoked in the context of using that particular term in the context of those particular objects. For example, 'front' may require objects to be closer to one another than 'back' because front sides of objects are typically the sides associated with the objects' function. This is certainly true for people, as 'front' corresponds to our direction of motion and the direction at which our perceptual apparati point (Clark 1973). It is also true of many artefacts, including televisions, microwave ovens, books, and clocks. Indeed, many objects define their 'front' side on the basis of it being the side with which one typically interacts (Fillmore 1971). As such, certain ranges of distances may be associated with 'front', much as certain ranges of distances are associated for terms such as 'near', with the particular value selected on the basis of characteristics of the objects being related. To assess this, we examined distance estimates as a function of spatial term, using an items analysis that held constant the objects being spatially related (i.e. comparing 'The squirrel is in front of the flower' to 'The squirrel is behind the flower'). The overall mean estimates as a function of term are shown in Figure 8.7. Estimates are clustered into groups, with group boundaries demarcated with a vertical line, and terms within a group receiving similar shading.

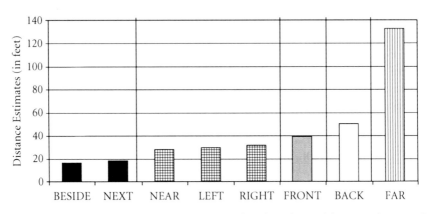

FIGURE 8.7. Mean distance estimates associated with each spatial term. Clusters of terms that are associated with similar distances are demarcated with a vertical line, with terms within a cluster receiving the same shading.

There are several important points to note. First, the mean distance associated with a given term should not be interpreted as corresponding to a fixed distance that the term conveys. Indeed, the means were obtained by averaging over objects, and the data in Figure 8.6 indicate that the particular objects being related significantly affect the particular value that is assigned. Rather, it is the pattern of rank ordering of the clusters of terms that we are interested in. It should be noted, however, that we have obtained independent estimates of the sizes of the located and reference objects, and we are currently using these in conjunction with the distance estimates as a means of assessing whether a term is likely to convey a range of distances that is linked to object size (i.e. expressing the distances in object units). Second, 'beside' and 'next' seem to suggest the smallest distance, and 'far' the largest. The distinction between 'beside' and 'next to' and 'near' is interesting, because these have been classified differently by Landau & Jackendoff (1993), with 'beside' and 'next to', but not 'near', conveying direction information in addition to distance information (see also Logan & Sadler 1996). How such differences arise as a consequence of the settings of different parameters of a reference frame is an interesting question, and raises the more general issue of interactions among the parameters. Third, 'near', 'left', and 'right' all seem to convey a similar distance. This is an interesting finding because 'near' is preferred over 'left' and 'right' owing to the potential ambiguity in the source of information assigning the orientation and direction for the latter terms (Mainwaring et al. 2003). The fact that these terms are used interchangeably would suggest that terms that convey a different distance (e.g. 'beside' or 'next') would not be considered viable alternatives, even though they are more similar to 'left' and 'right' by virtue of having a directional component. Fourth, the distances associated with 'front' are smaller than those associated with 'back', consistent with the idea that the function of an object is often associated with its front, and that successful interaction with the front may require a smaller distance between the objects. It would be interesting to see whether use of the term 'front' carries an implication that the objects are interacting or are about to interact. For example, contrast sentences (10) and (11).

(10) The squirrel is in front of the tree.
(11) The squirrel is behind the tree.

The question would be whether a person is more likely to infer a future interaction between the squirrel and the tree (e.g. the squirrel is about to climb the tree) in (10) than in (11). In summary, the pattern of clustering seems to suggest that spatial terms imply different ranges of distances.

## 8.5  Distance in 3D space

### 8.5.1  *A new methodology for examining distance in real space*

The examination of distance in the studies in sections 8.3 and 8.4 have occurred in somewhat limited contexts, with Carlson & van Deman (2004) assessing the encoding of distance between objects that are presented within a two dimensional display on a computer monitor, and Carlson & Covey (2005) assessing the distance that is inferred within conceptual space upon comprehending a linguistic description. Arguably, the more typical use of spatial language is describing the location of a target object with respect to co-present objects within an environment in 3D space. Accordingly, Carlson (forthcoming) describes a new methodology for examining how distance is encoded in 3D space. Specifically, participants were presented with a large (102 × 82 cm) uniform white board with a reference object in the center. The reference object was a cabinet (5 cm width × 7 cm length × 24 cm height) from a 'Barbie' dollhouse that was oriented to face the participant (see Figure 8.8). There were two versions of the cabinet, one with the door opening on the cabinet's left side (and on the right side of the viewer facing the cabinet, as in Figure 8.8), and one with the door opening on the cabinet's right side (and on the viewer's left). Extending out from the reference object toward the participant were 11 lines, as numbered in Figure 8.8. As described further below, the task of the participant was to make several distance judgements involving 'front' with respect to each

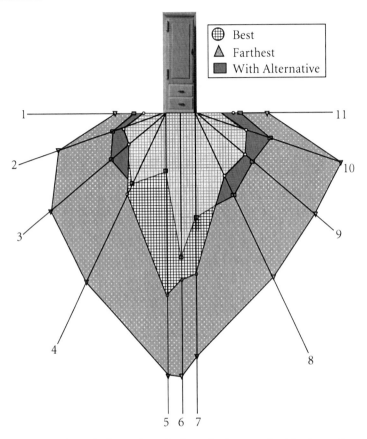

FIGURE 8.8. Locations designated as the best, farthest, and with alternative uses of front along 11 lines extending out from a dollhouse cabinet. See text for details.

of these lines. The lines were selected to correspond to particular regions of space around the cabinet.

Specifically, Logan & Sadler (1996) define a spatial template for a given spatial term as the space surrounding a reference object which is further divided into regions that indicate the acceptability of using the spatial term for describing the position of a located object within that region. The specific regions have been referred to as 'good', 'acceptable', and 'bad'. The 'good' region typically corresponds to the best use of a spatial term, and extends directly out from the relevant side of the reference object, such that the locations within the good region fall within the boundaries drawn by extending the edges of the relevant side into space. In Figure 8.8, lines 5–7 fall within the good region of the 'front' spatial template. The 'acceptable' region typically corresponds to permissible uses of the spatial term, and flanks the good region such that locations within the acceptable region fall outside the edges of the object but in the same direction as the good region. In Figure 8.8, lines 2–4 and 8–10 fall within the acceptable region of the 'front' spatial template. The 'bad' region typically corresponds to unacceptable uses of the spatial term, and extends in directions other than that indicated by the good region. In Figure 8.8, lines 1 and 11 fall within the bad region, extending to the viewer's left and right sides rather than to the front. As spatial templates are considered parameters of a reference frame (Carlson-Radvansky & Logan 1997), one can redefine the regions with respect to the axes of a reference frame, with the good region comprising locations along the relevant endpoint ('front') of the relevant axis ('front'/'back') with the reference frame axis running through this region, the acceptable region comprising locations flanking the relevant endpoint of the relevant axis, and the bad region comprising locations defined with respect to the other endpoint ('back') of the relevant axis ('front'/'back') or endpoints on the other axes.

Multiple lines were selected from within each of these regions. For the good region, line 6 extends from the center of the cabinet, corresponding to the object's visual center of mass. For other projective terms that require a reference frame (such as 'above'), the best definition of the term corresponds to locations that coincide with the reference object's center of mass, with a drop-off in the acceptability of the use of the term as one moves out to either side, but still within the good region (Regier & Carlson 2001). If participants judge 'front' in 3D space relative to the object's center of mass, then distance judgements associated with line 6 should differ systematically from judgements associated with lines 5 and 7, which should not differ. However, Carlson-Radvansky et al. (1999; see also Carlson & Kenny 2006) have shown that, in addition to defining terms relative to the center-of-mass, participants are also influenced by the location of functionally important parts of an object. They asked participants to place pictures of located objects above or below pictures of reference objects. The reference objects were shown from a sideways perspective, so that the center of mass of the object was dissociated from an important functional part that was located at the side of the object. For example, one of the stimulus pairs involved placing a tube of toothpaste above a toothbrush (presented in profile), with the bristles of the toothbrush shown off to the (viewer's) right, and the handle to the (viewer's) left. Placements were significantly biased away from the center of mass, toward the functional part. If functional information affects the conception of 3D space around the object, then judgements associated with the functional part should differ from judgements of other locations within the good region. Specifically, lines 5 and 7 are matched in distance away from line 6 at the center of the cabinet. However, for the cabinet with the door on the object's left (viewer's right), line 7 corresponds to locations in line with the functional part; for the cabinet with the door on the object's right (viewer's left), line 5 corresponds to locations in line with the functional part. A functional bias would correspond to a difference in judgement between lines 5 and 7, with the effect reversing across the two cabinets. For the acceptable region, lines extended out from the reference object at 22.5°, 45°, and 67.5°. Given that the acceptability within this region drops off as a function of angle deviation (Regier & Carlson 2001), use of multiple lines allowed us to assess the drop-off function as mapped onto 3D space. Finally, for the bad region, lines extended 90° to either side. These are baseline conditions that correspond to locations for which 'front' should not be used.

On each trial, a dowel was placed along one of the lines, and participants were asked to make three distance judgements pertaining to 'front'. In their initial judgement, they indicated the location on the dowel that corresponded to the best use of 'front'. Centimeters were marked along the side of the dowel facing away from the participant, and we noted this location. It was thought that this location would correspond most closely to the peak of the spatial template in an acceptability rating task (e.g. Hayward & Tarr 1995; Logan & Sadler 1996).

Participants then indicated how far away from the reference object they could move along the line and still use the term 'front'. This location was noted. This question was included for several reasons. First, it gave us a marker for the boundaries of the term. Intuitively, there seems to be a limit to the extension into space for which one might use a given term. For example, imagine two objects on an otherwise empty table, with one object in the middle and the other in front of it. Now imagine moving the object in front further and further away. At some point, 'front' becomes less acceptable, perhaps as the distance between the two objects becomes larger than the distance between the moving object and the edge of the table. At this point, it is likely that 'front' would be replaced with a description that related the object to the edge. Second, previous research measuring the use of spatial terms in restricted spaces (i.e. within a 2D computer display) have observed mixed effects of distance on acceptability judgements, with a distance effect observed with a larger range of distances (Carlson & van Deman 2004) but not a smaller range (Carlson & Logan 2001; Logan & Compton 1996). This paradigm allowed us to assess this question within a larger space. Finally, this measure enabled us to determine whether distance would vary as a function of the different types of region, and gave us a marker for the boundaries of the term.

In their final judgement, participants were asked whether an alternative spatial term would be preferred to describe the location at the far distance. If so, this alternative was noted, and as a follow-up, participants were asked to move along the line from that far location toward the reference object, stopping at the point at which 'front' became preferred over the alternative term that they had provided. With respect to the imaginary table example, this last placement would indicate the point at which 'front' remained preferred over an edge-based description. This question was included because it seemed likely that there could be a discrepancy in whether participants might consider use of the term permissible (i.e. the far location), but in actual use would prefer an alternative over this term. This question thus was an initial attempt at examining the impact of the presence of these alternative expressions on the space corresponding to 'front'.

Figure 8.8 shows average plots associated with these three measures. Distances are plotted at ¼ scale. The boxed region on the top corresponds to locations that were the best 'front', the largest region corresponds to the distances for which participants were willing to use 'front', and the dark intermediate region corresponds to the locations that participants moved back to in the case of alternative terms at the far locations. Note that not all participants had alternative terms, and so not all participants provided a distance for this third judgement. In addition, the likelihood of an alternative term varied as a function of line, with very few alternatives indicated for lines 5–7, and the use of an alternative increasing as a function of angle in the acceptable regions for lines 2–4 and 8–10. Note, too, that there was variability in the type of alternative, with some alternatives modifying 'front' ('almost front') and some alternatives switching to a different term ('left'). Thus, it is not clear that representing these alternatives collapsed over participants and collapsed over the type of alternative most accurately characterizes the impact of the alternative. For the present, we will focus on the best and farthest placements.

Several experiments were conducted using this methodology to examine how the space corresponding to 'front' was affected by factors including the type of reference frame used to define 'front', the location of the functional part, the type of term (comparing 'front' and 'back'), the conceptual size of the reference object (conceptual size: model-scale versus real-world, while equating absolute size), the addition and type of located object, and the presence of a distractor. We will discuss the results that compare 'front' and 'back', the location of the functional part, and the addition and type of a located object (see further Carlson, forthcoming).

## 8.5.2. *Initial findings*

### 8.5.2.1 *'Front'/'back' and the location of a functional part*   Figure 8.9, panel A, shows 'best' distances associated with 'front' and 'back' as a function of the location of the functional part, contrasting the cabinet with the door on the object's left to the cabinet with the door on the object's right. Different participants contributed to each condition. With respect to 'front' judgements, there is a clear effect of the location of the functional part. For the cabinet with the door on the object's left (viewer's right), distances associated with the functional line (line 7) were shorter than distances associated with the nonfunctional line (line 5). The same result was obtained for the cabinet with the door on the object's right (viewer's left), with distances associated with the functional line (line 5) shorter than distances associated with the nonfunctional line (line 7). Note the reversal in distances across these two lines as a function of the location of the functional part. This suggests that knowledge of how one would interact with the cabinet affected the distance associated with 'front'. The fact that the distances were smaller on the functional side is reminiscent of the smaller distances associated with 'front' in section 8.4 that we speculated as being do to functional interactions with the object's front side. In contrast, distances associated with lines 5 and 7 for the term 'back' did not differ, indicating instead that 'back' judgements may have been made with respect to the center of mass of the cabinet. The lack of an effect of door location was not due to participants not knowing how the door opened, because although the cabinet was facing away from the participant, its back was removed, and the door was left ajar so that participants could see how it opened. Rather, it is likely that there was no effect because the back side is not the side with which one typically interacts, and therefore functional information should not be relevant.

There is one potential caveat for the 'front' data in Figure 8.9, panel A. Participants in these conditions also made a series of judgements defining 'front' with respect to a viewer-centered reference frame, with some participants making the object-centered judgements first with a given cabinet, and others making the viewer-centered judgements first. Thus, these participants were exposed to both cabinets. Because the cabinets were identical except for the way in which the door opened, we expected that object-based responses to the second cabinet would be functionally biased because of the contrast between the two cabinets. In some sense, this is not problematic. Bub, Masson, & Bukach (2003) have shown that information about the function of an object is not automatically activated during identification. Thus, the contrast between the two cabinets may have been sufficient for activating functional knowledge, consequently resulting in a functional bias. Such a finding is still important, however, because there was nothing in the instructions to encourage

Panel A

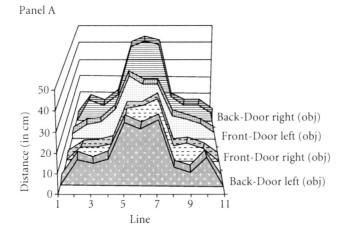

Panel B

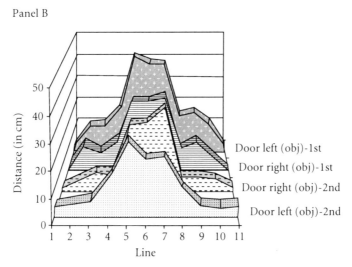

FIGURE 8.9. Panel A: Comparison of best locations as a function of term (front or back) and location of functional part (cabinet with door on object's left and cabinet with door on object's right). Panel B: Comparison of front placements as a function of cabinet, and whether locations were obtained in the first set of trials with exposure to only one location of the cabinet door (1st) or in a second set of trials with exposure to both door locations (2nd).

participants to pay attention to the door at any time. Thus, the extent to which participants brought this knowledge to bear on their distance judgements suggests that they deemed the information relevant. Figure 8.9, panel B, shows the 'front' data from panel A, as a function of whether the judgements were associated with participants' first set of trials (and thus, first experience with

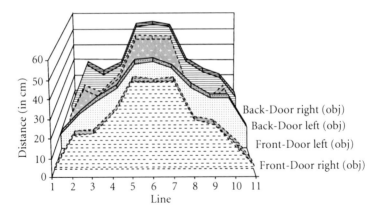

FIGURE 8.10. Comparison of farthest locations as a function of term (front or back) and location of functional part (cabinet with door on object's left and cabinet with door on object's right).

any of the cabinets) or with the second set of trials (and thus, a contrasting experience in which they had previously judged a cabinet with the door on the other side). Note the functional bias is present for both sets of data, although the effect is stronger in the second set of trials. This difference indicates an enhancement of functional information, presumably due to the contrast. The general presence of a functional bias indicates that information about the objects affects the configuration of 3D space, consistent with findings with other terms examining 2D space (e.g. Carlson-Radvansky et al. 1999).

Figure 8.10 shows the data for 'front' and 'back' for the farthest placements. Overall, no effect of function is observed at this distance, although there is a trend (seen in Figure 8.8) for 'front' with the cabinet with the door on the object's left. This also makes sense given a functional interaction explanation: at this far distance, interaction with the door of the cabinet is not possible; accordingly, the impact of the location of the functional part should not play a role.

8.5.2.2 *The addition and type of located object*    Figure 8.11 shows the impact of including a located object, either a Barbie doll or a small dog (Barbie's pet, at a scale consistent with the dollhouse cabinet and Barbie). Different participants placed different objects at the 'best' (Panel A) and 'farthest' (Panel B) locations around the cabinet with the door on the object's right. Consider first the 'best' locations shown in Panel A. We have included the 'best' distance data for 'front' with the cabinet with the door on the right from the experiments described in section 8.5.2.1, in which participants indicated locations without placing a located object. This is listed as the 'No Object' condition.

Panel A

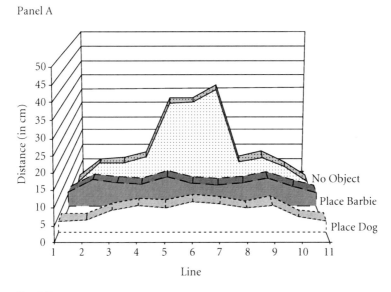

Panel B

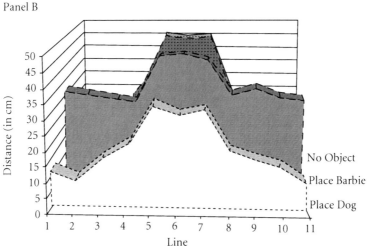

FIGURE 8.11. Best front locations associated with placing a located object, either a Barbie or a dog. Locations from the comparable condition in which locations were indicated along the dowel are included (No object) for comparison.

Comparison with the both the 'Place Barbie' and 'Place Dog' conditions reveals a significant impact due to the addition of the located object. Rather than locations peaking in the good region (consistent with many other studies examining spatial templates, including Carlson-Radvansky & Logan 1997; Hayward & Tarr 1995; Logan & Sadler 1996), the locations are flat across the lines associated with the good and acceptable regions. They are also considerably closer to the reference object. This suggests that participants may have been defining front with respect to the interaction between the objects and the cabinet. Consistent with this interpretation, notice that the distances associated with the (smaller) dog are smaller than those associated with Barbie. This effect is reminiscent of the object size effect discussed in section 8.4. Indeed, it may be profitable to interpret these data along the lines developed in section 8.4. The experiment in which participants judged 'front' by indicating locations along the dowel can be viewed as establishing a range of possible 'front' values. A particular value from this range was then selected on the basis of the particular located object that was selected, in such a way as to maximize their potential for interaction. For the dog, this distance was a bit closer to the reference object than it was for Barbie. As shown in Panel B, these effects also hold for the farthest distance data, with locations associated with Barbie and the dog both within the range established by the 'No Object' condition, and with locations associated with the dog closer to the reference object than locations associated with Barbie. These results demonstrate a significant impact of the identity of the located object and the way in which it interacts with the reference object on the conceptualization of space associated with 'front' around the reference object.

## 8.6  Conclusions and implications

The findings based on three different methodologies converge on the conclusion that distance is an essential component to the processing of spatial language, both for terms that explicitly convey distance as part of their definition (such as 'near') and for terms that do not explicitly convey distance. The data in section 8.3 suggest that during the processing of spatial language, distance between the objects is encoded and retained, presumably within a parameter of a reference frame. This distance setting seems to operate at the level of the reference frame, applying to terms on all axes. The data from section 8.4 suggest that spatial terms that do not explicitly convey a distance may be associated with a range of distances from which a particular value is selected as a function of the particular objects being spatially related. The data from section 8.5 offer converging evidence for this point from a different methodology, and demonstrate that these effects operate within 3D space. Thus, distance is fundamental to spatial language.

In many ways this is not a surprising conclusion. As attested by the many chapters of this volume, space is foundational to cognition. Thus, showing that an aspect of space (distance) is computed during processing (spatial) language seems both trivial and obvious. One reason that this conclusion has been somewhat overlooked is that studies of language and space have largely examined the mapping in the direction of asking how language is assigned onto space. Within this approach, the linguistic term and its features are identified, and then associations between these features and characteristics of space are made. What this approach leaves out is all of the other dimensions of space that could be relevant to the processing of spatial language but that may not be explicitly marked within the linguistic term itself. Take distance as an example. Because terms like 'above' and 'front' do not convey distance as part of their definitions, the assumption has been made that distance is therefore irrelevant to their processing. However, this one-way mapping of linguistic features of the term onto space also misses out on the additional cognitive work that is operating during apprehension of spatial language, including the perception of the objects, allocation of attention to the objects, construction of a situation model that also represents the goals of the utterance vis-à-vis the speaker and the listener, and so on. All of these aspects are part and parcel of processing spatial language, and features derived thereof can also be mapped onto space. With respect to distance, attention moves from one object to the other during the processing of spatial language (Logan 1995); this then is a possible mechanism for defining distance. That distance then becomes associated with the use of the spatial term indicates that such aspects of space are attended to and deemed relevant at a level that is more cognitive than linguistic. In this way, this mapping of space onto language is especially compatible with the embodied cognition approach to language comprehension, most particularly, Zwaan's (2004) immersed experiencer model and more generally the idea of grounding language in action (Glenberg 1997; Glenberg & Kaschak 2002) and in perceptual simulations (Barsalou 1999; Pecher, Zeelenberg, & Barsalou 2003). Within this general view, our cognition is tied to our experiences in the world, and the understanding of any given cognitive process (including the use of spatial language) should acknowledge the way in which the world itself may infiltrate this process. In this chapter we have reviewed evidence suggesting that one component of the world (distance in space) significantly impacts spatial language. This is but one example of this general point, as attested by the other chapters in this volume.

# Section III
# Using Space to Ground Language

The chapters in Section I of this book looked at how abstract thought is guided and constrained by the experience of living in space as a spatial being. Section II focused in on spatial cognition itself. The chapters in this section return to the grounding of high-level cognition, specifically language. Language has appeared in various guises in almost every chapter in this book, especially in the chapters by Carlson, Clark, Mix, Ramscar et al., Spencer et al., and Spivey et al., but in this section the relationship between space and language is tackled head-on. The three papers in this section examine three quite distinct aspects of space and the perception of space that are crucial to the use and the learning of language.

This relation between language and space has been studied much like the relation between language and every other cognitive domain (Bloom et al. 1999; Hickmann & Robert 2006; Levinson 2003; Levinson & Wilkins 2006): How does language partition the domain? To what extent is the partitioning constrained by universals, whether linguistic or extra-linguistic? How different are the partitions found in different languages, and what implications do these differences have for non-linguistic cognition? How do language learners 'get into' the system built into the language they are learning? These studies are often forced to confront questions about the abstractness of spatial representation that we have seen in the first two sections of this book because language, by its very nature, imposes a level of abstractness on a cognitive domain.

One of the chapters in this section, Cannon and Cohen, falls to some extent within this domain of enquiry (as do the chapters by Carlson and by Lipinski et al.). Cannon and Cohen are interested in how we perceive, classify, and represent the dynamics of our environment (e.g. object motions) and how these representations, in turn, serve to ground verb meanings. They explore the hypothesis that there exists a 'semantic core' of perceptual primitives out of which verb meanings are constructed (including, perhaps, even seemingly 'intentional' distinctions such as 'avoid' vs. 'pursue'). This core is inherently spatial, dynamic, and interactive. That is, Cannon and Cohen go beyond the usual investigations of language about space to argue in effect that language, even when it is not explicitly spatial, starts with space. In this sense, they are in agreement with a body of work originating in linguistics (some that they refer to: Lakoff 1987; Talmy 1988), as well as recent behavioral studies of the semantics of verbs reported on in Spivey et al.'s chapter in this book, that takes spatial cognition to be behind much of language, metaphorically if not directly. In emphasizing the role of motion, their work also resonates with that reported by Ramscar et al. in their chapter in Section I of this book. Both chapters make the case that the perception of motion is fundamental to symbolic cognition.

There is another, less obvious way that space and language interact, in relation to how referents for new words are established. As Quine (1960) pointed out, learners cannot discover the meaning of a new word without first determining what thing or event in the world the speaker means to address. The challenge of explaining how language learners solve this problem has led some to posit various innate constraints on word learning such as 'mutual exclusivity' and the 'whole-object constraint' (Markman, 1989). Alternatives to such innatist accounts start from the well-known fact that most of the language addressed to young children (like Quine's original 'Gavagai' example, as described in Yu et al.) is in the 'here and now', that is, that it concerns events and objects that the child can observe and may even be a part of. Even though such utterances may be not be *about* space, they are inherently spatial in the sense that they are *deictic*: they *point* to things that are located in the observable world. Given this constraint, however, there are still two problems that have to be solved: the child must use clues to figure out which object or event is associated with which word, and must keep track of the objects and the words long enough to associate them with one another.

The other two chapters in this section offer novel accounts of how children solve these two problems. Each builds on a body of work on non-linguistic spatial cognition. Yu and Ballard start with the literature on how infants learn to be sensitive to adults' gaze direction and how toddlers learn to be aware of adults' referential intentions. They then set out to show, first in experiments with adults, then in a sophisticated computational model, how information about shifts in the gaze of speakers as they describe scenes in the here-and-now actually does facilitate word learning. The implication is that language comprehension builds on the listener's internal representations of the intricate relation between the speaker's body and the objects in the environment, and that these representations are crucial in language learning. Recent work in social robotics (Breazeal et al. 2006) shows how these insights are crucial to the construction of robots that solve Quine's problem in interaction with human teachers.

Smith starts with two well-known phenomena—the perseveration that characterizes infant motor behavior and is best exemplified in the A-not-B phenomenon, and the association of objects in the visual field to locations in egocentric space in short-term memory. She shows how children use spatial short-term memory to maintain links between words they have heard and particular positions in body-centered space. This happens even in the absence of temporal contiguity between the word and its referent; it is enough for the word to be associated with the place that then comes to *stand in for* the referent. Smith argues that this sort of perseveration and perseveration in the A-not-B task are reflections of the same underlying phenomenon. That is, the same mechanisms that facilitate action also facilitate binding of labels to objects. Something like the perseveration that Smith shows in young children is fundamental to the way in which discourse is structured in signed languages. Signers quite literally place referents in particular positions in signing space, where they can later be referred to; the locations come to stand for the referents themselves (Friedman 1975). Smith's argument is also consistent with the picture painted in Clark's chapter in this book: space as resource to simplify online processing demands.

In sum, all three of these chapters converge on the centrality of space to language: language *needs* space. Comprehending and producing language is spatial in at least three senses. (1) Speakers and listeners rely, concretely or abstractly, on spatial categories. This is true not only because speakers are listeners are *in* space, but also because space, in particular motion, matters to them (see also the introduction to Section I). (2) Speakers direct their bodies at the things they are referring to and listeners make use of this information in interpreting sentences (and learning the meanings of unfamiliar words). (3) Speakers and listeners use space to offload some of the memory demands by deictically assigning things to places.

The converse—that space needs language—has not been a major theme of this book, but it is worth mentioning here because of the renewed attention it is receiving. The view that the very spatial nature of human existence bears on even the most abstract forms of cognition, that abstract cognition is in fact facilitated by space, that language in particular relies on space in a number of ways—none of this presupposes that processes operate in the opposite direction, that linguistic categories influence the experience of space. Interest in the possibility of such processes goes back to the proposals of Whorf and Sapir regarding what is now often called 'linguistic determinism' (Sapir & Mandelbaum 1949; Whorf & Carroll 1964). While some of the best-known early empirical work designed to disprove (or prove) this hypothesis concerned color (Berlin & Kay 1969), the investigation of spatial language and its possible influence on spatial perception has been a frequent preoccupation since then. While much of this research remains controversial, and while it may ultimately be impossible to disentangle linguistic from *cultural* influences, the existence of some of these effects seems now undeniable. Particularly compelling are effects on perspective taking (Emmorey, Klima, & Hickok 1998; Levinson 2001) and on the categorization of simple dynamic events (Bowerman & Choi 2003).

The possibility of language-specific influences on spatial perception bears on many, if not all, of the issues of concern in this book. It has three general sorts of implication. First, computational models may need to be modified to allow for the influence of language on spatial representations. It is not clear how either the DFT model described in Lipinski et al.'s chapter or the hybrid model described by Yu and Ballard could, in their current forms, incorporate such effects. For example, Yu and Ballard's model learns its visual and linguistic representations independently; there is no way for the linguistic categories to modify the visual categories. Second, one should be cautious in drawing conclusions about 'spatial cognition' on the basis of data from speakers of a single language. For example, this caution would apply to spatial representations underlying verbal memory, imagery, and online comprehension (Spivey et al.'s chapter); to the use of language and space as a unified resource for reduction of complexity (Clark's chapter); and to the role of distance in language processing (Carlson's chapter). Third, as Cannon and Cohen suggest in their chapter and as already tackled in part by Boroditsky in work related to her chapter with Ramscar and Matlock (Boroditsky 2001), cross-linguistic, cross-cultural research is called for.

# 9

## Objects in Space and Mind: From Reaching to Words

### LINDA B. SMITH AND LARISSA K. SAMUELSON

Traditionally, we think of knowledge as enduring structures in the head—separated from the body, the physical world, and the sensorimotor processes with which we bring that knowledge to bear on the world. Indeed, as formidable a theorist as Jean Piaget (1963) saw cognitive development as the progressive differentiation of intelligence from sensorimotor processes. Advanced forms of cognition were those that were separate from the here-and-now of perceiving and acting. This chapter offers a new take on this idea by considering how sensorimotor processes, specifically the body's orientation and readiness to act in space, help to connect cognitive contents, enabling the system to transcend space and time. In so doing, we sketch a vision in which cognition is connected to sensorimotor functions, and thus to the spatial structure of the body's interactions in the world, but *builds* from this a more abstract cognition removed from the specifics of the here and now.

### 9.1 The A-not-B error

Piaget (1963) defined the object concept as the belief that objects persist in space and time and do so independently of one's own perceptual and motor contact with them. He measured infants' 'object concept' in a simple object-hiding task in which the experimenter hides an enticing toy under a lid at location A. After a delay (typically 3 to 5 seconds), the infant is allowed to reach—and most infants do reach correctly—to A and retrieve the toy. This A-location trial is repeated several times. Then, there is the crucial switch trial: the experimenter hides the object at a new location, B, as the infant watches. But after the delay, if the infant is 8 to 10 months old, the infant will make a characteristic 'error', the so-called A-not-B error. They reach not to where they saw the object disappear, but back to A, where they found the object previously. Infants older than 12 months of age do not typically perseverate but search correctly on the crucial B trials (see Wellman 1986). Piaget suggested that this pattern indicated that older infants but not younger ones know that objects can exist independently of their own actions. There has, of course, been much debate about this conclusion and many relevant experiments pursuing a variety of alternatives (Acredolo 1979; Baillargeon 1993; Bremner 1978; Diamond 1998; Munakata 1998; Spelke & Hespos 2001), including that infants much younger than those who fail the traditional A-not-B task do—in other tasks—represent the persistence of the object beyond their own perceptual contact.

In this context of divergent views on the phenomenon, Smith, Thelen, and colleagues (Smith, Thelen, Titzer, & McLin 1999; Thelen & Smith 1994; Thelen, Schoner, Scheier, & Smith 2001; Spencer, Smith, & Thelen 2001) sought to understand infants' *behavior* in the task. At the behavioral level, the task is about keeping track of an object's location in space and reaching to that right location; hence, infants' failures and successes in the task may be understood in terms of the processes that underlie visually guided reaching. From the perspective of visually guided reaching, the key components of the task can be described analyzed as illustrated in Figure 9.1. The infant *watches* a series of events, the toy being put into a hiding location and then covered with a lid. From this, the infant must formulate *a motor plan* to reach and must *maintain this plan* over the delay, and then execute the plan. This motor plan, which necessary in *any* account of infants' actual performance in this task, *in and of itself* may implement a 'belief'—a stability in the system—that objects persist in space and time.

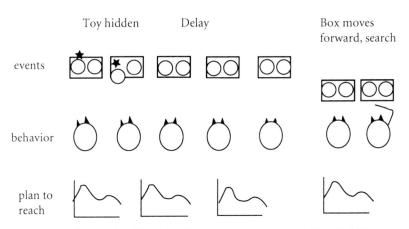

FIGURE 9.1. A task analysis of the A not B error, depicting a typical A-side hiding event. The box and hiding wells constitute the continually present visual input. The specific or transient input consists of the hiding of the toy in the A well. A delay is imposed between hiding and allowing the infant to search. During these events, the infant looks at the objects in view, remembers the cued location, and undertakes a planning process leading to the activation of reach parameters, followed by reaching itself.

Thelen et al. (2001) developed a formal account enabling them to understand how the error might emerge in the dynamics of the processes that form and maintain a reaching plan. The theory is illustrated in schematic form in Plate 4. The larger figure illustrates the activation that is a plan to move the hand and arm in a certain direction. Three dimensions define this motor planning field. The x-axis indicates the spatial direction of the reach, to the right or left. The y-axis indicates the activation strength; presumably this must pass some threshold in order for a reach to be actually executed. The z-axis is time. All mental events occur in real time, with rise times, durations, and decay times. In brief, the activation in the field that is the plan to reach evolves in time as a function of the sensory events, memory, and the field's own internal dynamics.

According to theory, activation in this field is driven by three inputs to the field. The first is the continually present sensory activation due to the two covers on the table. These drive activation (perhaps below a reaching threshold) to those two locations because there is something to reach to at those locations. The second input is the hiding event that instigates a rise in activation at the time and location of the hiding of the object. It is this activation from this specific input that must be maintained over the delay if the infant is to reach correctly on B trials. The third input is the longer-term memory of the previous reaches, which can perturb the evolving activation in the field, pulling it in the direction of previous reaches.

Plate 5 shows results from simulations of the model. Plate 5A illustrates the evolution of activation in the hypothesized motor planning field on the very

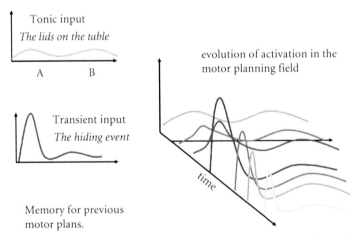

4. An overview of the dynamic field model of the A not B error. Activation in the motor planning field is driven by the tonic input of the hiding locations, the transient hiding event, and the memories of prior reaches. This figure shows a sustained activation to a hiding event on the left side despite recent memories of reaching to the right, that is a nonperseverative response. For improved image quality and colour representation see Plate 4.

first A trial. Before the infant has seen any object hidden, there is low activation in the field at both the A and B locations that is generated from the perceptual input of the two hiding covers. As the experimenter directs attention to the A location by hiding the toy, that perceived event produces high transient activation at A. The field evolves and maintains a planned reaching direction to A. This evolution of a sustained activation peak that can drive a reach even after a delay, even when the object is hidden, is a consequence of the self-sustaining properties of the dynamic field. Briefly, the points within a field provide input to one another such that a highly activated point will exert a strong inhibitory influence over the points around it, allowing an activation to be maintained in the absence of external input.

This is a dynamic plan to reach, and is continuously informed by the sensory input and continuously driven by that input. But the activation in the field is also driven—and constrained—by memories of previous reaches. Thus, at the second A trial, there is increased activation at site A because of the previous activity there. This combines with the hiding cue to produce a second reach to A. Over many trials to A, a strong memory of previous actions builds up. Each trial embeds the history of previous trials. Plate 5B illustrates the consequence of this on the critical B trial. The experimenter provides a strong cue to B by hiding the object there. But as that cue decays, the lingering memory of the actions at A begin to dominate the field, and indeed, over the course of the delay through the self-organizing properties of the field itself activation shifts back to the habitual, A side. The model predicts that the error is time-dependent: there is a brief period immediately after the hiding event when infants should search correctly, and past research shows that without a delay, they do (Wellman 1986).

The model makes a number of additional predictions that have been tested in a variety of experiments (see Thelen et al. 2001; Clearfield, Dineva, Smith, Fiedrich, & Thelen 2007). Because it is continuously tied to the immediate input, visual events at hiding, at the moment of the reach, and indeed even after the reach has begun, can drive a different solution and push the reach to A or to B. Indeed, simulations from the model can be used to design experimental manipulations that cause 8- to 10-month-olds to search correctly on B trials and that cause 2- to 3-year-olds to make the error (Spencer et al. 2001). These effects are achieved by changing the delay, by heightening or lessening the attention-grabbing properties of the covers or the hiding event, and by increasing and decreasing the number of prior reaches to A (Diedrich, Highlands, Spahr, Thelen, & Smith 2001; Smith, Thelen, Titzer, & McLin 1999). All these effects show how the motor plan is dynamically connected to sensory events and to motor memories. Because one can make the error come and go in these ways over a broad range of ages (from 8 to

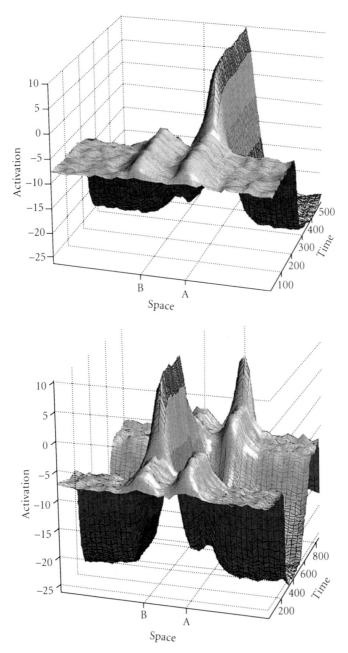

5. (A) The time evolution of activation in the planning field on the first A trial. The activation rises as the object is hidden and due to self-organizing properties in the field is sustained during the delay. (B) The time evolution of activation in the planning field on the first B trial. There is heightened activation at A prior to the hiding event due to memory for prior reaches. As the object is hidden at B, activation rises at B, but as this transient event ends, due the memory properties of the field, activation. For improved image quality and colour representation see Plate 5.

30 months), we know that the relevant processes cannot be tightly tied to one developmental period. Instead, they may reflect general processes that govern spatially directed action across development. Thelen et al.'s model explicitly incorporates this idea by showing how the model yields seemingly qualitatively distinct patterns—perseveration, non-perseveration—through small changes in the parameters.

## 9.2  Representation close to the sensorimotor surface

The processes that underlie the behavior—the activations in the dynamic field—are specifically conceptualized as motor plans, plans that take the hand from its current location to the target. Memories for previous reaches—the memories that create the perseverative error—are also motor plans. Conceptualizing the processes and memories in this way leads to new predictions about the role of the body and its position in space. To be effective, motor plans *must* be tied to specific body position to plan a movement *from the current position* to the intended target. By this account, the processes that create the error should depend on the current postural state of the body. If this is so, then the A-not-B is a truly sensorimotor form of intelligence, just as Piaget suggested.

This prediction has been borne out in a series of experiments. The key result is this: distorting the body's posture appears to erase the memory for prior reaches and thus the cause of perseveration, leading to correct searches on the B trials (Smith, Clearfield, Diedrich, & Thelen, in preparation; Smith et al. 1999). For example, in one experiment, infants were in one posture (e.g. sitting) on A trials and then shifted to another posture (e.g. standing) on B trials. This posture shift between A and B trials (but not other kinds of distraction) caused even 8- and 10-month-old infants to search *correctly*, supporting the proposal that the relevant memory is a motor plan. More specifically, the results tell us that the relevant memories are in the coordinates of the body's position and coupled to those current coordinates such that these memories are not activated unless the body is in that posture. This makes sense: Motor plans—if the executed action is to be effective—must be tied to the *current* relation of the body to the physical world, and the relevant motor memories for any current task are those compatible with the body's current position.

In many ways, these results—and the conceptualization of the relevant processes as motor plans—fit well with Piaget's (1963) original conceptualization of the error: as an inability to represent objects independently of their sensorimotor interactions with objects. As Piaget noted, it is as if the mental object, and its location, are inseparable from bodily action. One conclusion some might want to take from these studies is that the A-not-B error has little to do with object representation *per se* and is instead about interfering motor habits. Certainly, there may well be other systems that remember objects and that are not so tied to sensory motor processes. We will return to this idea later in this chapter. However, at present, we want to emphasize that motor plans are *one* system that provide a means of representing non-present objects, and given the nature of motor plans, these representation (or memories) bind the object to a body-defined location.

Such representations may be considered, as they were by Piaget, to be a limitation of an immature cognitive system. But the proposed mechanisms are also hypothesized to be general mechanisms of visually guided reaching, and thus not specific to immature systems. In the next section, we consider older infants and a task structurally similar to the A-not-B task but in which perseveration, after a fashion, yields the successful mapping of a name to a non-present referent. We show further that the processes that enable this mapping are also a kind of sensorimotor intelligence.

## 9.3  From reaching to words

Infants appear to keep track of objects in the A-not-B task by forming a plan for action, a plan of how to move the hand to reach the object. Because the motor plan necessarily specifies a target, it binds the object to that location. Given this binding, activation of one component gives rise to the other. Thus, in the A-not-B task, activation of the object and the goal to reach yields the prior motor plan and a reach to A rather than to B. But this is just one possible consequence of the binding of an object to an action plan. Activation of the action plan (and past location) should work to call forth the memory of the non-present object. In this section, we describe a series of just-completed studies that show how the same processes that create the A-not-B error also create coherence in other tasks, enabling the coherent connection of an immediate sensory event to the *right* contents in the just previous past. The new task context concerns how young children (18 to 24 months of age) map names to referents when those names and referents are separated in time and occur in a stream of events with multiple objects and multiple shifts in attention among those objects. The experiments use an ingenious task created by Baldwin (1993) to study early word learning, but one which is, in many ways, a variant of the classic A-not-B task.

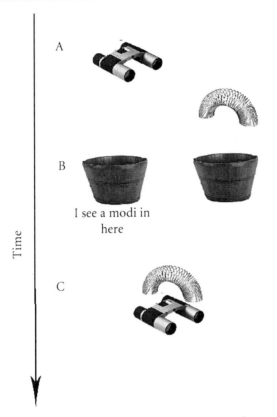

A

B

I see a modi in
here

C

Time

FIGURE 9.2. Events in the Baldwin task. See text for further clarification.

The stream of events in Baldwin's task is illustrated in Figure 9.2. The experimenter sits before a child at a table, and (a) presents the child with first one un-named object on one side of midline and then with a second un-named object on the other side. Out of sight of the child, the two objects are then put into containers and the two containers (b) are placed on the table. The experimenter looks into one container and says, 'I see a modi in here.' The experimenter does not show the child the object in the container. Later the objects are retrieved from the containers, presented in a new location (c), and the child is asked which one is 'a modi'. Notice that the name and the object were never jointly experienced. The key question is whether the child can join the object name to the right object even thought that object was not in view when the name was heard. Baldwin showed that young children could do this, taking the name to refer to the *unseen* object that had been in the bucket at the same time the name was offered.

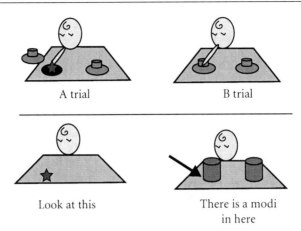

FIGURE 9.3. An illustration of two time steps in the A-not-B task and the Baldwin task. In the A-not-B task, children repeatedly reach and look to locations to interact with objects at location A, causing motor planning memory biased to location, and in this way binding the object to location A. In the Baldwin task, children repeatedly reach and look to locations to interact with objects. This causes objects—through remembered motor plans and attentional plans—to be bound to locations; children can then use this binding of objects to locations to link a name to a non-present object.

Figure 9.3 illustrates the A-not-B task and the Baldwin task showing their similar surface structures. In the A-not-B task, children turn to look at and reach to an object at a particular location. By hypothesis, this binds the object to an action plan. Subsequently, the goal of reaching for that object activates that plan and the old target location. In the Baldwin task, children turn attention to two different objects at two different locations, and again by hypothesis bind these objects to spatially specific action plans of looking and reaching. If this analysis is correct, then just as the goal to reach for the one object in the A-not-B task may call up the spatially specific action plan, then so could activation of the action plan—a look to a specific location—activate the memory of the object associated with that plan. In the Baldwin task, this would lead to the child remembering the right object at the moment the experimenter offered that name. Could this parallel account of perseveration in the A-not-B task and successful mapping of a name to a non-present thing in the Baldwin task possibly be right? Is success in the Baldwin task due to the same processes that yield failure in the A-not-B task? A series of recently completed experiments support this idea (Smith 2009).

The first experiment in the series sought to show that a common direction of spatial attention was critical to children's successful linking of the name to the object in Baldwin's task. As in the original Baldwin study, the participants were young children, 18 to 14 months of age. The experiment replicated the original Baldwin method, and in a second condition sought to disrupt children's word-object mappings by disrupting the link between the object and a single direction of attention and action, making spatial location more variable and less predictive of specific objects. The method in this new condition is the same as that in Figure 9.2A *except* that the two objects were each, prior to the naming even, presented once on each side, that is once on the right and once on the left. Thus each object was associated with both directions of attention prior to the naming event. If children keep track of objects through their action plans, and if they link a heard name to a remembered object by common direction of looking, then this inconsistency in an object's location prior to naming should disrupt the mapping the name to the thing in the bucket. The results indicate that it does. When each object was consistently linked to one direction of attention, Baldwin's original result was replicated and the children chose the target object—the object in the bucket during the naming event—73% of the time. In constrast, when the objects were not consistently linked to one direction of attention, children chose the target object only 46% of the time, which did not differ from chance. These results indicate that children are learning *in the task* about the relation between objects and their locations. A stronger link between direction of attention and an object makes direction of attention a better index to the memory of that object.

One difference between the A-not-B task and the Baldwin task is the goal in the A-not-B task is to reach to and obtain the toy. Thus, it seems reasonable that the relevant memory—a memory represents the existence of the object when it is out of view—might be tied to a motor plan. But is this reasonable for the Baldwin task? Is turning attention to a location also rightly conceptualized as a motor plan? As Allport (1990) put it, attention is for action; we attend to locations in preparation for possible action, and actions are performed toward objects within the focus of attention. Accordingly, as one test of this conceptualization, we asked if it is the body's direction of attention—and not a specific location in space—that enables children to access the right object when the experimenter provides the name. If our analysis of the direction of attention as a motor plan for orienting the body in space is correct, then one should be able to activate the memory for one object or the other simply by shifting the child's orientation in one direction or the other. For example, pulling attention generally to the left during the naming event should activate memories for the object seen on the left, and then the name should be linked to that object. All aspects of the procedure were identical to that illustrated in Figure 9.2 except, at the moment of naming, there was just an empty table top, no buckets, no hidden objects. With the experimenter looking straight into the child's eyes, but with one hand held an arm-stretch to the side, she clicked her fingers and said 'Modi, Modi, Modi.' The clicking fingers and the outstretched hand directed children's attention to that side, so that when they heard the name, they were looking to one side or the other. This directional orientation influenced their mapping of the name to the non-present object. Children chose the object spatially linked to the clicking fingers 68% of the time. This suggests that the direction of bodily attention is bound to the object. These results also highlight how the solution to connecting experiences separated in time may be found in the child's active *physical* engagement in the task.

If the relevant links in the Baldwin task between objects and locations are through motor plans and thus close to the sensory surface, are they also disrupted by shifts in the body's postures? The bodily direction of attention, like a reach, is necessarily egocentric. Just how one shifts one's direction of gaze or turns one's body to bring an object into view depends on the body's *current* position in relation to that object. Accordingly, in this next experiment, we again altered children's posture—from sitting low and close to the table such that looks right and left (and reaches left and right) required lateral moves, to one in which the child was standing on the edge of the table of itself, so that the child looked down with a bird's eye view of the locations. The method was the original Baldwin task with buckets as in Figure 9.2. More specifically, in one condition, children sat (as in the previous experiments) when the objects were first presented, unnamed, and each associated with opposing directions of attention. Then during the naming event, the child was stood up and remained standing through out the procedure. If the memory of previously experienced objects is strongly linked to plans for action, then this posture shift—by causing a reorganization of that plan—should disrupt the memory. Recall that in the A-not-B task, this disruption caused a lessening of perseveration and more correct responding. The prediction, here, is that the posture shift during the naming event should disrupt retrieval of the target object, and children should not be successful in mapping the name to the object. The full experiment included four conditions: sit-sit, with a visual distraction before the naming event; stand-stand, with a visual distraction before the naming event; sit-stand; and stand-sit. In the visual distraction (no posture shift) conditions, children chose the target (spatially linked) object on 70% of the test trials; in the posture-shift conditions, they did so 50% of the time, performing at chance. These results strongly suggest that the memory for previously experienced objects is in or indexed through processes tightly tied to the body's current orientation in space. These are memories in spatial coordinates tied to the body's position. Again, this is a form of sensorimotor intelligence in that the relevant processes appear to close to the sensorimotor surface.

These processes that give rise to children's successful mapping of a name to a physically non-present object also appear fundamentally the same as those that lead even younger children to reach perseveratively in Piaget's classic A-not-B task. In one case these processes create *coherence* in the cognitive system, appropriately linking the right object in the just previous past to the immediate input (the naming event), enabling the child to keep track of objects as referents in a conversation and stream of events with attention shifting from one location to another. In the other, the processes create *perseveration*, an inappropriate sticking to the past when the immediate cues call for a shift to a new response. If the cognitive coherence in the Baldwin task is a form of 'good' perseveration—one that yields a positive outcome—then one should be able to alter the Baldwin task in a way that these same processes will yield an inappropriate 'sticking' to the just previous past. Accordingly, we attempted to create an A-not-B effect within the Baldwin task.

The reasoning behind this version of the experimental task is this: If attentional direction activates memories associated with that bodily orientation, these activated memories should compete with mapping a name to a physically present object if it is at the same place. Prior to naming, objects were presented four times, one always on one side of the table and the other always on the other side, in order to build up a strong link between a direction of bodily attention and a seen object. During naming, the experimenter showed the child one object, pointed to it and named it *but did so with the object (and thus the child's attention) at the side associated with the other object.* This sets up a possible competition between the just previously experienced object at this location and the present one that was being named. Given this procedure, children selected the named object only 42% of the time, despite the fact that it was in view and pointed to when named. Clearly, the prior experience of seeing one object in a particular location disrupted linking the name to a physically present object at that same location. This pattern strongly supports the idea that the direction of attention selects and activates memories from the just previous past, creating in this case interference but also—in the standard Baldwin task—enabling very young children to bind events in the present to those in the just previous past.

These results point to the power of sensorimotor intelligence: how being in a body and connected to a physical world, is part and parcel of human cognition. They fit with emerging ideas about 'cheap' solutions (see O'Regan & Noë 2001), about how higher cognitive ends may be realized through the continuous coupling of the mind *through the body* to the physical world. Young children's solution to the Baldwin task is 'cheap' in the sense that it does not require any additional processes other than those that must already be in place for perceiving and physically acting in the world.

These results also fit a growing literature on 'deictic pointers' (Ballard et al. 1997), and is one strong example of how sensorimotor behaviors—where one looks, what one sees, where one acts—create coherence in our cognition system, binding together related cognitive contents and keeping them separate from other distinct contents. One experimental task that shows this is the 'Hollywood Squares' experiments of Richardson and Spivey (2000). People were presented at different times with four different videos, each from a distinct spatial location. Later, with no videos present, the subjects were asked about the content of those videos. Eye-tracking cameras recorded where people looked when answering these questions, and the results showed that they systematically looked in the direction where the relevant information had been previously presented.

The strong link between the bodily orientation of attention and the contents of thought is also evident in everyday behavior. People routinely and apparently unconsciously gesture with one hand when speaking of one protagonist in a story and gesture with the other hand when speaking of a different protagonist. In this way, by hand gestures and direction of attention, they link separate events in a story to the same individual. Children's bodily solution to the Baldwin task may be another example of this general phenomenon and the embodied nature of spatial working memory.

## 9.4 Transcending space and time

The present analysis sees children's success in the Baldwin task (as well as infants' perseveration in the A-not-B task) as a form of sensorimotor intelligence, an intelligence bound to the here-and-now of perceiving and acting. However, not all aspects of children's performances in the Baldwin task fit this idea. Critically, once children map the name to the object, their knowledge of that mapping does not appear to be spatially fixed. This is seen at testing when the child is provided with the name and asked to choose to which of two objects it refers. In all the tasks, the objects are presented at a new location and, as shown in Figure 9.2, are overlapping and on top of one another. Thus, the direction of prior attention associated with the name and with the object cannot be used to determine the intended target. The processes that children use to learn the name and the processes that are available to them once the name has been learned appear to be different. When children form the link between the object and the name, they appear to use the spatial orientation of their body to retrieve from memory the non-present object. But once this mapping is made, they apparently no longer need any spatial correspondence with past experiences with the object for the name to direct attention to the object. The course of events in this experiment is thus reminiscent of Piaget's (1963) grand theory of cognition, in which he saw development as a progression from sensorimotor intelligence to representations that were freed from the constraints of the here-and-now. In the Baldwin task, children use sensorimotor processes to keep track of things in space and mind, but naming brings a new index to memory that does involve the body's disposition to act.

With John Spencer and Gregor Schöner, we are currently working on an extension of the dynamic field model to explain this. Plate 6 illustrates the general idea. At the top are sensorimotor fields, one for objects and the spatial direction of attention (and action) and one for sounds and the spatial direction of attention (and action). Within this theory, these are sensorimotor fields because they are driven by and continuously coupled to the sensory input, and because they specify an action plan for directing attention. As in the original dynamic field model of the A-not-B error, these fields also are driven by memories of their own recent activation. What is new is that these sensory fields are also coupled to each other and to a new kind of field, an association field that has, itself, no direct sensory input and is not a plan for action. Instead, the word-object association field only has only inputs from the object-space field

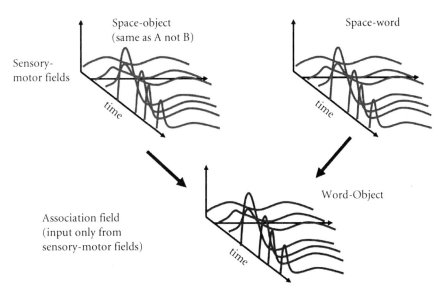

6. Illustration of how two sensory-motor fields representing attention and planned action to objects in space and to sounds in space may be coupled and feed into an association field that maps words to objects without represesenting the spatial links of those words and objects. For improved image quality and colour representation see Plate 6.

and the word-space field, and represents the associations between words and objects in a manner that is unconnected to the spatial context of experience. It is, in this formulation, the association field that frees the mapping of word and object to spatially directed action plans.

This general idea is also similar to proposals by Damasio (1989) and Simmons & Barsalou (2003) about the origin of higher cognition in multi-modal sensory processes and their associations to each other. Figure 9.4 illustrates the general idea. Consistent with well-established ideas in neuroscience, there are modality-specific and feature-specific areas of sensory and motor (and emotional) representations. These feed into a a hierarchical system of association areas. At lower levels, association areas exist for specific modalities, capturing feature states within a single modality. At higher levels, cross-modal association areas integrate feature states across modalities and give rise to higher-order regularities that are more abstract, and that transcend modality-specific representations, but that are nonetheless built from them. The role of

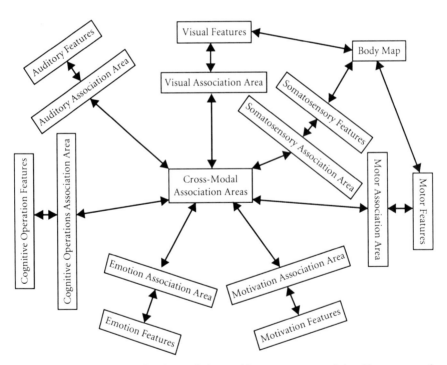

FIGURE 9.4. A conceptualization of the architecture proposed by Simmons and Barsalou, in which sensory and motor areas specific to specific modalities and features interact and create multimodal association areas.

the association areas, then, is to capture modality-specific states for later representational use. These association areas integrate feature activations across modalities forming higher-level representations that transcend the specific sensorimotor systems. The extension of the dynamic field model and the proposal of a word-object association field fits this general idea of a multi-modal outside-in architecture.

The evidence reviewed here on the role of sensorimotor processes in the A-not-B task and in the Baldwin task suggest the following:

(1) The processes that create perseverative reaching and the A-not-B error are a truly sensorimotor form of intelligence, embedded in the processes that form and maintain motor plans.

(2) These same processes—processes that create what would seem to be a deficiency in infant cognition—also play a positive role in enabling young word learners to keep track of objects and cognitive contents over time and to coherently bind them to each other.

(3) This sensorimotor intelligence is a stepping stone to representations distinct from those sensorimotor representations.

The remainder of this chapter considers several broader implications of these ideas.

## 9.5  Space as an index to objects

The A-not-B error is typically conceptualized as a signature marker of a cognitive or neural immaturity, and also, in mature individuals, as a marker of neural damage (e.g. Diamond 1990a; 1990b). This is because the error is highly predictive of frontal-lobe functioning, and is related to the executive control functions that coordinate multiple neural systems in the service of a task and that enable flexible shifting in those coordinations. The proposal we offer below is *not* at odds with this well-established understanding, but offers a complementary perspective as to why there is such a strong link between objects and their location in the first place—a link so strong that a developing (or damaged) executive control system finds it so difficult to override. We propose that this link between object and location is a fundamental aspect of the human cognitive system, one seen with particular clarity in the developing infant. In this way, the A-not-B error is revealing not just about the development of executive control but also about the fundamentally spatial organization of object representations in working memory.

The idea that objects are indexed in working memory by space is one with a long history in cognitive psychology. One example is Posner's (1980) space-based account of visual selection which saw attention as a lingering spotlight over a spatially circumscribed area. Among the considerable findings consistent with this view is the result that reaction times to detect a target are faster when the target falls in the same location as a preceding spatial cue (see Posner 1980). More recent research also builds a strong case for object-based visual selection (e.g. Chun & Wolfe 2001; Humphreys & Riddoch 2003; Luck & Vogel 1997; Moore, Yantis, & Vaughan 1998).

Other evidence suggests that location information plays a particularly key role in working memory. Both infants and adults appear to implicitly learn an object's location when they attend to that object and then subsequently use that location information to refind the object. (Spivey, Tyler, Richardson, & Young 2000; Richardson & Kirkham 2004). Moreover, as noted earlier, if the object is not physically present, participants use location to retrieve information about it, physically turning the body's sensors toward the location in which the to-be-retrieved event occurred—a move that in turn fosters successful retrieval (Spivey et al. 2000). Richardson and Kirkham (2004) refer to this phenomenon as 'spatial indexing', and suggest that objects and associated events are stored together with memory addressed by spatial location. Spatially organized attention also plays an important role in Kahneman & Treisman's (1992) theory of object files. According to this account, focal attention to a location is the glue that binds features to make an object. More specifically, attention to a location activates the features *at that location* and integrates them into a temporary object representation in working memory, forming a spatially indexed object file that contains the object properties.

In brief, although there are many detailed disputes in this literature, there is a general consensus that within task contexts, objects are indexed in working memory by their location. The A-not-B error seems to be fundamentally about these processes, and might be best understood within this literature. Indeed, the immaturity of executive control systems may allow us to see more clearly the early embodied nature of attention and working memory.

From this perspective, the A-not-B task and the Baldwin task add to our understanding of the location-based indexing of objects in two ways. First, they show how this indexing is, at least early in development, realized close to the sensory surface, so close that body's posture and momentary positioning matter. Second, they show how this spatial indexing plays a significant role in young children's ability to keep track of words and referents when their occurrences are separated in time. In this way, the body's momentary disposition space helps create coherence in the stream of thought. In a specific task context, over the course of a conversation, with many different contents and shifts of attention, one can nonetheless coherently connect the events separated in time by the body's orientation in space (see also Ballard et al. 1997).

## 9.6 Attention from the outside in?

Richardson and Kirkham's notion of spatial indexing has its origins in Ullman's (1984) proposal about 'deictic primitives'. Ullman introduced the term to refer to transient pointers used to mark aspects of a visual scene. In computer vision systems, deictic pointers reduce the need to store information about a scene by allowing the system frequent and guided access to the relevant sensory input (e.g. Lesperance & Levesque 1990). The power of deictic pointers comes from the need for only a relatively small number of these pointers, each bound to a task-relevant location in a scene. Pointers bound to a location are formed and maintained only so long as they are relevant to the immediate task. Since only the pointers need to be stored, rather than what they point at, there is a significant reduction in memory demands.

Pointers *can* be thought of as internal and symbolic, with little direct relation the physical act of pointing in space. And certainly, internal pointers—not linked to the sensorimotor system—may well exist in the cognitive system. But findings that adults shift their eye gaze to the past source of to-be-remembered information, and findings about the role of the body's orientation in space in the A-not-B and Baldwin tasks, raise the possibility that internal pointing systems might be deeply related to the body's physical positioning of sensors and effectors, reflecting the spatial constraints of a physical body in a physical world. At the very least, an internal attentional system must be compatible with bodily forms of attending.

In this context, we wonder if at least some attentional proceses, particularly attention shifting and executive control, might progress from the outside in. For infants and very young children, attention—and attention switching—may be linked to physical action and require movement of the whole body to the proper orientation in space, in this way unsticking attention—shifting attention given new task goals—by new spatial coordinates for action. With increasing development, smaller and more subtle movements may suffice—perhaps just a shift in eye gaze. Later, internal simulations of movement, never outwardly realized, may be sufficient. At present this is conjecture, but a developmentally intriguing one, as greater attentional control may emerge because of the tuning of internal systems by external actions.

This idea that internal cognitive operations mirror corresponding physical action is an instance of the isomorphism proposed by Shepard (1975; Shepard & Chipman 1970) between physical properties and their mental representation.

Shepard & Chipman (1970) distinguished between first-order and second-order isomorphisms. In first-order isomorphism, characteristics of a physical stimulus are literally present in the representation of that stimulus: for example, the representation of a larger object might involve more neurons than the representation of a smaller object. Shepard & Chipman dismissed this form of isomorphism as unlikely. In second-order isomorphism, characteristics of a physical stimulus are preserved in a more analogical or functional way: for example, things that have physically similar properties might be represented near each other in some neural or perceptual space. One potential example of a second-order isomorphism related to that proposed here in relation to attention is the mental rotation of three-dimensional objects (Shepard & Metzler 1971). When participants are asked to imagine a rotating object, the temporal and spatial properties of these rotations mirror the properties of actually physically rotating the object. Barsalou (2005) suggested that these kinds of isomorphism may be understood as internal simulations that use—without actual outward action—the same sensorimotor processes that would execute the physical action to create internal dynamic representations. In this way, external bodily direction of attention to and action on objects may set up dynamic internal representations and attentional mechanisms.

## 9.7  An object concept through multiple forms of representation

Piaget (1963) considered the perserverative searches of infants to indicate the lack of an object concept—in the sense of an inability to represent the object independently of one's own actions on it. The present results seem consistent with this view. The internal representations that give rise to perseveration in the A-not-B task do represent the object as enduring when it is out of sight, but they do so through sensorimotor processs tied very much to the momentary disposition of the body. There are a variety of other measures of the object concept, measures that do not involve reaching but only looking, in which infants do seem to represent the persistence of objects (e.g. Baillargeon 1993). The role of the body in these representations has not been systematically examined, but given the nature of the task, these representations may be not strongly linked to action plans. This does not mean that such representations are not also realized close to the sensory surface in modality-specific representations at the outer ring of processes illustrated in Figure 9.4. A complex heterogeneous system such as the human cognitive system is likely to have many mutually redundant systems of representation.

The interesting idea—and one already put forth (though in somewhat different forms) by such developmental giants as Piaget (1963), Vygotsky

(1986), and Bruner (1990)—is that the these sensorimotor representations generate—through their associations with each other and perhaps critically with language—new forms of abstract representations freed from the here-and-now of sensorimotor experience. Children's performances in the Baldwin task clearly make this two-edged point: cognition is grounded to here-and-now through the body, and the body's physical position in space appears to play a strong role in structuring cognition; yet from these processes emerge other, more abstract forms of representation distinct from those processes.

# 10

## The Role of the Body in Infant Language Learning

CHEN YU AND DANA H. BALLARD

Recent studies have suggested both that infants use extensive knowledge about the world in language learning and that much of that knowledge is communicated through the intentions of the mother. Furthermore, those intentions are embodied through a body language consisting of a repertoire of cues, the principal cue being eye gaze. While experiments show that such cues are used, they have not quantified their value. We show in a series of three related studies that intentional cues encoded in body movement can provide very specific gains to language learning. A computational model is developed based on machine learning techniques, such as expectation maximization, which can identify sound patterns of individual words from continuous speech using non-linguistic contextual information and employ body movements as deictic references to discover word-meaning associations.

It is quite obvious that thinking without a living body is impossible. But it has been more difficult to appreciate that our bodies are an interface that represents the world and influences all the ways we have of thinking about it. The modern statement of this view is due to Merleau-Ponty (1968).

If we started with sheet music, everything that Mozart ever wrote would fit on one compact disc, but of course it could never be played without the instruments. They serve as a 'body' to interpret the musical code. In the same way human bodies are remarkable computational devices, shaped by evolution to handle enormous amounts of computation. We usually are unaware of this computation, as it is manifested as the ability to direct our eyes to objects of interest in the world or to direct our body to approach one of those objects and pick it up.

Such skills seem effortless, yet so far no robot has come close to being able to duplicate them satisfactorily. If the musculoskeletal system is the orchestra in our metaphor, vision is the conductor. Eye movements are tirelessly made at the rate of an average of three per second to potential targets in the visual surround. In making tea we pre-fixate the objects in the tea-making plan before each step (Land, Mennie, & Rusted 1999). In a racquet sport we fixate the bounce point of the ball to plan our return shot (Land & Mcleod 2000). Since the good resolution in the human eye resides in a central one degree, our ability to maneuver this small ball throughout a volume of over a million potential fixation points in a three-dimensional world is all the more impressive. Nonetheless we do it. We have to do it, for our musculoskeletal system is designed with springs, and the successful manipulation of objects is dependent on the ability to preset the tension and damping of those springs just before they are needed. This interplay of fixation and manipulation is a central feature of primate behavior, and preceded language, but it contains its own implicit syntax. The 'I' is the agent. The body's motions are the verbs and the fixations of objects are nouns. In infant language learning, the 'you' is the caregiver, a source of approval and other reward. Given this introduction, the reader is primed for our central premise: that the body's natural 'language' can serve and in fact did serve as a scaffold for the development of spoken language.

## 10.1 Early language learning

Infant language learning is a marvelous achievement. Starting from scratch, infants gradually acquire a vocabulary and grammar. Although this process develops throughout childhood, the crucial steps occur early in development. By the age of 3, most children have incorporated the rudiments of grammar and are rapidly growing their vocabulary. Perhaps most impressively, they are able to do this from the unprocessed audio stream which is rife with ambiguity. Exactly how they accomplish this remains uncertain. It has been conjectured that it may be possible to do this by bootstrapping from correlations in the audio stream, and indeed recent experimental evidence demonstrates that the cognitive system is sensitive to features of the input (e.g. occurrence statistics). Among others, Saffran, Newport, & Aslin (1996) showed that 8-month-old infants are able to find word boundaries in an artificial language only based on statistical regularities. Later studies (Saffran, Johnson, Aslin, & Newport 1999) demonstrated that infants are also sensitive to transitional probabilities over tone sequences, suggesting that this statistical learning mechanism is more general than the one dedicated solely to processing linguistic data. The mechanisms may include not only associative processes but also algebraic-like computations to learn grammatical structures (rules). The recent work in Pena, Bonatti, Nespor, & Mehler (2002) showed that silent gaps in a continuous speech stream can cause language learners to switch from one computation to another.

In addition to word segmentation and syntax, the other important issue in language acquisition is how humans learn the meanings of words to establish a word-to-world mapping. A common conjecture of lexical learning is that children map sounds to meanings by seeing an object while hearing an auditory word form. The most popular mechanism of this word learning process is *associationism*. Richards & Goldfarb (1986) proposed that children come to know the meaning of a word through repeatedly associating the verbal label with their experience at the time that the label is used. Smith (2000) argued that word learning is initially a process in which children's attention is captured by objects or actions that are the most salient in their environment, and then they associate it with some acoustic pattern spoken by an adult. This approach has been criticized on the grounds that it does not provide a clear explanation about how infants map a word to a potential infinity of referents when the word is heard, which is termed *reference uncertainty* by Quine (1960). Quine presented the following puzzle to theorists of language learning: Imagine that you are a stranger in a strange land with no knowledge of the language or customs. A native says 'Gavagai' while pointing at a rabbit in the distance. How can you determine the intended referent? Quine offered this puzzle as an example of the indeterminacy of translation. Given any word-event pairing, there are in fact an infinite number of possible intended meanings—ranging from the rabbit as a whole, to its color, fur, parts, or activity. But Quine's example also includes a powerful psychological link that does rule out at least some possible meanings—pointing. The native through his body's disposition in space narrows the range of relevant perceptual information. Although not solving the indeterminacy problem, pointing (1) provides an explicit link between the word and location in space and in so doing (2) constrains the range of intended meanings. Thus, auditory correlations in themselves are unlikely be the whole story of language learning, as studies show that children use prodigious amounts of information about the world in the language process, and indeed this knowledge develops in a way that is coupled to the development of grammar (Gleitman 1990).

A large portion of this knowledge about the world is communicated through the mother's animated social interactions with the child. The mother uses many different signaling cues such as hand signals, touching, eye gaze, and intonation to emphasize language aspects. Furthermore, we know that infants are sensitive to such cues from studies such as Baldwin et al. (1996), Bloom (2000), and Tomasello (2000); but can we quantify the advantages that they offer? In this chapter, we report on three computational and experimental studies that show a striking advantage of social cues as communicated by the body. First, a computational analysis of the CHILDES database is presented. This experiment not only introduces a formal statistical model of word-to-world mapping but also shows the role of non-linguistic cues in word learning. The second experiment uses adults learning a second language to study gaze and head cues in both speech segmentation and word-meaning association. In the third experiment, we propose and implement a computational model that is able to discover spoken words from continuous speech and associate them with their perceptually grounded meaning. Similarly to infants, the simulated learner spots word-meaning pairs from unprocessed multisensory signals collected in everyday contexts and utilizes body cues as deictic (pointing) reference to address the reference uncertainty problem.

## 10.2 Experiment 1: statistical word learning

The first of our studies uses mother-infant interactions from the CHILDES database (MacWhinney & Snow 1985). These tapes contain simultaneous audio and video data wherein a mother introduces her child to a succession of toys stored in a nearby box. The following transcript from the database is representative of the mother's descriptions of one of the toys, in this case Big Bird from the television series *Sesame Street*:

hey look over here see the birdie
see the birdie
oh yes yes I know
let's see
you want to hold the bird you want hold this

The usefulness of non-linguistic cues has its critics, and this example shows why. In this kind of natural interaction, the vocabulary is rich and varied and the central item, Big Bird, is far from the most frequent word. Furthermore the tape shows numerous body language cues that are not coincident with the Big Bird utterance. This complex but perfectly natural situation can be easily quantified by plotting a histogram of word frequency for an extended sequence that includes several toys, as shown in the first column of Figure 10.1. None of the key toy items makes it into the top 15 items of the list. An elementary idea for improving the ranking of key words assumes that the infants are able to weight the toy utterances more by taking advantage of the approximately coincident body cues. For instance, the utterances that were generated when the infant's gaze was fixated on the toys by following the mother's gaze have more weights than the ones the young child just looked at when not paying attention to what the mother said. We examined the transcript and re-weighted the words according to how much they were emphasized by such cues, but, as the second column in Figure 10.1 shows, this strategy does little to help.

What *is* helpful is to partition the toy sequences (contextual information when the speech was produced) into intervals where within each interval a single toy or small number of co-occurring toys is the central subject or meaning, and then to categorize spoken utterances using the contextual bins labeled by different toys. Associating meanings (toys etc.) with words (toy names etc.) can be viewed as the problem of identifying word correspondences between English and a 'meaning language', given that the data of these two languages in parallel. With this perspective, a technique from machine translation can address the correspondence problem (Brown, Pietra, Pietra, & Mercer 1993). We apply the idea of Expectation Maximization (EM) (Dempster, Laird, & Rubin 1977) as the learning algorithm. Briefly speaking, the algorithm assumes that word-meaning pairs are some hidden factors underneath the observations which consist of spoken words and extralinguistic contexts. Thus, association probabilities are not directly observable, but they somehow determine the observations because spoken language are produced based on caregivers' lexical knowledge. Therefore, the objective of language learners or computational models is to figure out the values of association probabilities so that they can increase the chance of obtaining the observations. Correct word-meaning pairs are those which can maximize the likelihood of the observations in natural interactions. We argue that this strategy is an effective one that young language learners may apply during early word learning. They tend to guess most reasonable and most co-occurring word-meaning pairs based on the observations from different contexts.

The general setting is as follows: suppose we have a word set $X = \{w_1, w_2, \ldots, w_N\}$ and a referent set $Y = \{m_1, m_2, \ldots, m_M\}$, where N is the number of words and M is the number of meanings (toys, etc.). Let S be the number of learning situations. All word data are in a set $X = \{(S_w^{(s)}, S_m^{(s)}), 1 \leq s \leq S\}$, where for each learning situation, $S_w^{(s)}$ consists of r words $w_{u(1)}, w_{u(2)}, \ldots, w_{u(r)}$, and $u(i)$ can be selected from 1 to N. Similarly, the corresponding contextual information $S_m^{(s)}$ in that *m*th learning situation include *l* possible meanings $m_{v(1)}$, $m_{v(2)}, \ldots, m_{v(l)}$ and the value of $v(j)$ is from 1 to M. Assume that every word $w_n$ can be associated with a meaning $m_m$. Given a data set $X$, the task is to maximize the likelihood of generating the 'meaning' streams given English descriptions:

$$P\left(S_m^{(1)}, S_m^{(2)}, \ldots, S_m^{(S)} \mid S_w^{(1)}, S_w^{(2)}, \ldots, S_w^{(S)}\right) = \prod_{s=1}^{s} \sum_a p(S_m^{(1)}, a \mid S_w^{(S)})$$

The technical descriptions can be found in Yu & Ballard (2004). Figure 10.1 shows that this algorithm strikingly improves the probability of the toy vocabulary. 65% of words are associated with correct meanings, such as the word *hat* paired with the meaning 'hat' (third column) and the word *book* paired with the meaning 'book' (fourth column). In addition, all the toy words are in the top three of the corresponding objects (columns). Note that the object 'ring' (the fifth column) seems to relate to multiple words. That is because in the video clips, the mothers introduced to the children to a set of rings with different colors. Therefore, they spent significantly more time on the object 'ring', and consequently many words co-occur more frequently with the meaning 'ring' compared with other meanings.

In contrast to previous models of cross-situational learning (e.g. Siskind 1996) that are based inference rules and logic learning, our proposed model is based on probabilistic learning and is able to explicitly represent and estimate the association probabilities of all the co-occurring word-meaning pairs in the training data. Moreover, this formal model of statistical word learning provides a probabilistic framework to study the role of other factors and constraints in word learning, such as social cues and syntactic constraints. The results demonstrate the potential value of this mechanism—how multimodal correlations may be sufficient for learning words and their meanings. We also want to note two major assumptions in this computational study: (1) infants can segment words from continuous speech; and (2) they can partition the interaction intervals based on the focal toy. Our computational model described in Experiment 3 uses unprocessed multisensory data to associate spoken words with their perceptually grounded meanings, and demonstrate that body cues play a key role in grounding language in sensorimotor experiences. In the following, we will first present an experimental study that provides empirical support for our argument of the role of body cues, and then in section 10.4 describe the grounded model which provides a mechanistic explanation of how it works.

## 10.3  Experiment 2: deictic body cues in human simulation

A major advance in recent developmental research has been the documentation of the powerful role of social-interactional cues in guiding the infants learning and in linking the linguistic stream to objects and events in the world. Studies (e.g. Baldwin 1993; Baldwin et al. 1996; Tomasello 2000; Bloom 2000, Woodward & Guajardo 2002) have shown that there is much information in social interaction, and that young learners are highly sensitive to that information. Butterworth (1991) showed that even by 6 months of age, infants demonstrate sensitivities to social cues, such as monitoring and following another's

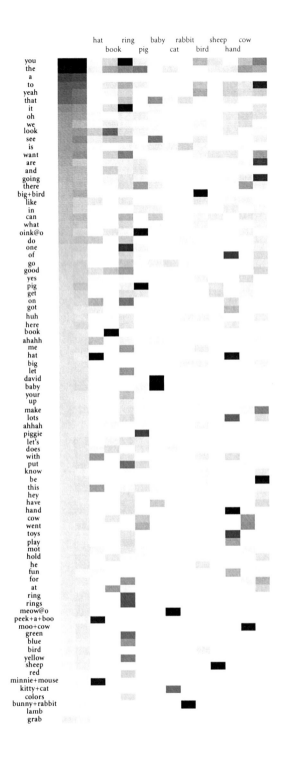

gaze, although infants' understanding of the implications of gaze or pointing does not emerge until approximately 12 months of age. Based on this evidence, Bloom (2000) suggested that children's word learning in the second year of life actually draws extensively on their understanding of the thoughts of speakers. Similarly, Tomasello (2000) showed that infants are able to determine adults' referential intentions in complex interactive situations, and he concluded that the understanding of intentions, as a key social cognitive skill, is the very foundation on which language acquisition is built. These claims have been supported by experiments in which young children were able to figure out what adults were intending to refer to by speech. For example, Baldwin et al. (1996) proposed that 13-month-old infants give special weight to the cues of indexing the speaker's gaze when determining the reference of a novel label. Their experiments showed that infants established a stable link between the novel label and the target toy only when that label was uttered by an adult who concurrently directed their attention (as indexed by gaze) toward the target. Such a stable mapping was not established when the label was uttered by a speaker who showed no signs of attention to the target toy, even if the object appeared at the same time that the label was uttered and the speaker was touching the object. However, there is an alternate understanding of these findings to the proposals of 'mind reading'. Smith (2000) has suggested that these results may be understood in terms of the child's learning of correlations among actions, gestures, and words of the mature speaker, and intended referents. Samuelson & Smith (2000) argued that construing the problem in this way does not so much 'explain away' notions of 'mind reading' as ground those notions to the perceptual cues available in the real-time task that infants must solve. Further, grounding such notions as 'referential intent' and 'mind reading' in correlations among words, objects, and the coordinated actions of speakers and listeners provides a potential window into more conceptual understandings of referential intent. In relation to this idea, Baldwin & Baird (2001) proposed that humans gradually develop the skill of mind reading so

FIGURE 10.1. Word-like unit segmentation. First column: the histogram of word frequency from Rollins's video data in the CHILDES database shows that the most frequent words are not the central topic meanings. Second column: weighting the frequency count with cues improves the situation only slightly. Remaining columns: the results of statistical word learning to build word-to-world mappings. The row is a list of words and the column is a list of meanings. Each cell is the association probability of a specific word-meaning pair. White color means low probability while dark means high probability. In our model, spoken utterances are categorized into several bins that correspond to temporally co-occurring attentional objects. The EM algorithm discounts words that appear in several bins, allowing the correct word-meaning associations to have high probability.

that ultimately they care little about the surface behaviors of others' dynamic action, but instead focus on discerning underlying intentions based on a generative knowledge system.

In light of this, our second experiment documents the power of the body's disposition in space in helping language learning, and attempts to ask more directly if body cues are in fact helpful for both speech segmentation and word-meaning association, which are two cruxes in early language learning. As in Quine's example, the subjects are adults presented with a foreign word and a complex scene, and the task is to determine the meaning of the word. The experiment uses eye gaze rather pointing as the explicit from word to world. Using adults is only an indirect way to explore infant language learning. The adults being exposed to a new language have explicit knowledge about English grammar that is unavailable to infants, but at the same time do not have the plasticity of infant learners. Nonetheless, it has been argued that adult learning can still be a useful model (Gillette, Gleitman, Gleitman, & Lederer 1999). Certainly, if adults could not use body cues it would be an argument against their use in the infant model, but it turns out that the cues are very helpful.

### 10.3.1 *Data*

We use English-speaking adult subjects who are asked to listen to an experimenter reading a children storybook in Mandarin Chinese. The Mandarin is read in a natural tone similar to a caregiver describing the book to a child, and with no attempts to partition the connected speech into segmented words as was done in the first study. The reader is a native speaker of Mandarin describing in his own words the story shown in a picture book entitled 'I went walking' (Williams & Vivas 1989). The book is for 1–3-year-old children, and the story is about a young child who goes for a walk and encounters several familiar, friendly animals. For each page of the book, the speaker saw a picture and uttered verbal descriptions. Plate 7 shows visual stimuli in three learning conditions. In one condition, Audio only, the speaker's reading served as the stimulus training materials. In a second condition, Audio + Book, the Audio portion along with a video of the book as each page was turned served as the training material. In the third condition, Head and Eyes Cues, the audio portion, a video of the book as each page was turned, and a marker that showed where on the page the speaker was looking at each moment in time in the reading, served as the training material. In the audio-visual condition, the video was recorded from a fixed camera behind the speaker to capture a view of the picture book while the auditory signal was also presented. In the eye-head-cued condition, the video was recorded from a head-mounted camera to provide a dynamic first-person view. Furthermore, an eye tracker was utilized to track the time-course of the speaker's eye movements and gaze positions. These gaze positions were indicated by a cursor that was superimposed on the video of the book to indicate where the speaker was looking from moment to moment. Subjects were divided into three groups: audio-visual, eye-head-cued, and audio-only. The 27 subjects were randomly assigned to these three training conditions. Each listened (watched) the training material five times.

### 10.3.2 *Testing*

Testing differed somewhat for the three groups. All groups received a segmentation test: subjects heard two sounds and were asked to select one that they thought was a word but not a multi-word phrase or some subset of a word. They were given as much time as they wanted to answer each question. There were 18 trials. Only subjects in the audio-visual and eye-head-cued training conditions received the second test. The second test was used to evaluate knowledge of lexical items learned from the video (thus the audio-only group was excluded from this test). The images of 12 objects in the picture book were displayed on a computer monitor at the same time. Subjects heard one isolated spoken word for each question and were asked to select an answer from 13 choices (12 objects and also an option 'none of the above').

### 10.3.3 *Results*

Figure 10.2 shows the average percentage correct on the two tests. In the speech segmentation test, a single-factor ANOVA revealed a significant main effect of the three conditions $F(2; 24) = 23.52$; $p < 0:001$. Post hoc tests showed that subjects gave significantly more correct answers in the eye-head-cued condition (M = 80.6%; SD = 8.3%) than in the audiovisual condition (M = 65.4%; SD = 6.6%; $t(16) = 4.89$; $p < 0:001$). Performance in the audio-only condition did not differ from chance (M = 51.1%; SD = 11.7%). Subjects in this condition reported that they just guessed because they did not acquire any linguistic knowledge of Mandarin Chinese by listening to the fluent speech for 15 minutes without any visual context. Therefore, they were not asked to do the second test. For the word learning test, performance in the eye-head-cued condition was much better than in the audio-visual condition ($t(16) = 8.11$; $p < 0:0001$). Note also that performance in the audio-visual condition was above chance ($t(8) = 3.49$; $p < 0.005$, one-sample t tests).

The results show the importance of explicit cues to the direction of attention of the speaker, and suggest that this information importantly disambiguates potential meanings. This finding goes beyond the claims by Baldwin (1993) and Tomasello (2000) that referential intent as evidenced in gaze affects word learning. Our results suggest that information about the speaker's attention, a

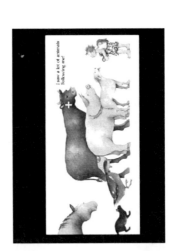

7. The snapshots when the speaker uttered "the cow is looking at the little boy" in Mandarin. Left: no non-speech information in audio-only condition. center: a snapshot from the fixed camera. Right: a snapshot from a head-mounted camera with the current gaze position (the white cross). For improved image quality and colour representation see Plate 7.

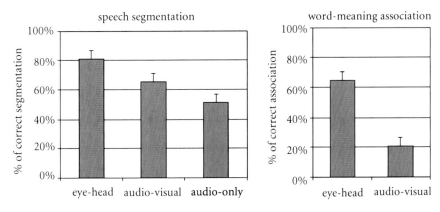

FIGURE 10.2. The mean percentages of correct answers in tests

social cue, not only plays a role in high-level learning and cognition but also influences the learning and the computation at the sensory level.

To quantitatively evaluate the difference between the information available in the audiovisual and eye-head-cued conditions, the eye-head-cued video record was analyzed on a frame-by-frame basis to obtain the time of initiation and termination of each eye movement, the location of the fixations, and the beginning and the end of spoken words. These detailed records formed the basis of the summary statistics described below. The total number of eye fixations was 612. Among them, 506 eye fixations were directed to the objects referred to in the speech stream (84.3% of all the fixations). Thus, the speaker looked almost exclusively at the objects that were being talked about while reading from the picture book. The speaker uttered 1,019 spoken words, and 116 of them were object names of pictures in the book. A straightforward hypothesis about the difference in information between the eye-head-cued and audiovisual conditions is that subjects had access to the fact that spoken words and eye movements are closely locked in time. If this temporal synchrony between words and body movements (eye gaze) were present in the eye-head-cued condition (but not in the audio-visual condition), it could explain the superior performance on both tests in the eye-head-cued condition. For instance, if the onset of spoken words were always 300 msec. after saccades, then subjects could simply find the words based on this delay interval. To analyze this possible correlation, we examined the time relationship of eye fixation and speech production. We first spotted the key words (object names) from transcripts and labeled the start times of these spoken words in the video record. Next, the eye fixations of the corresponding objects, which are closest in time

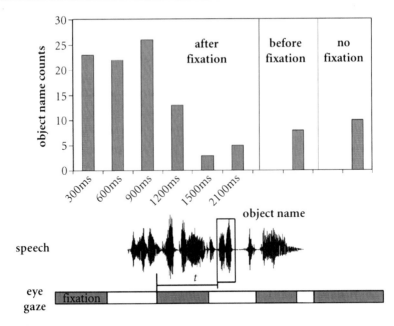

FIGURE 10.3. The level of synchrony between eye movement and speech production. Most spoken object names were produced after eye fixations and some of them were uttered before eye fixations. Occasionally, the speaker did not look at the objects at all when he referred to them in speech. Thus, there is no perfect synchrony between eye movement and speech production.

to the onsets of those words, were found. Then, for each word, we computed the time difference between the onset of each eye fixation and the start of the word. A histogram of this temporal relation is plotted to illustrate the level of synchrony between gaze on the target object and speech production. As shown in Figure 10.3, most eye movements preceded the corresponding onset of the word in the speech production, and occasionally (around 7%) the onset of the closest eye fixations occurred after speech production. Also, 9% of object names were produced when the speaker was not fixating on the corresponding objects. Thus, if the learner is sensitive to this predictive role for gaze-contingent co-occurrence between visual object and speech sound, it could account for the superior performance by subjects in the eye-head-cued condition on tests of both speech segmentation and word-meaning association. In the following study, we describe a computational model which is also able to use the information encoded by this dynamic correspondence to learn words. We also note here two important limitations of this experimental study: (1) the learners are adults and not children; (2) we marked the direction of eye gaze on the page; the learner did not have to figure it out. Still, the study demonstrates the potential importance of these cues in real-time learning.

## 10.4  Grounding spoken language in sensorimotor experience

The Mandarin learning experiment shows conclusively that eye gaze is a big help in retaining vocabulary information in a new language, but does not address the issue of the internal mechanism and provide a complete picture of early language learning. Thus, we want to know not only that learners use body cues but also *how* they do so in terms of the real-time processes in the real-time tasks in which authentic language learning must take place. We want to study learners' sensitivities to social cues that are conveyed through time-locked intentional body movements in natural contexts. In light of this, the last study introduces a computational model that learns lexical items from raw multisensory signals to closely resemble the difficulties infants face in language acquisition, and attempts to show how gaze and body cues can be of help in discovering the words from the raw audio stream and associating them with their perceptually grounded meanings.

The value of this approach is highlighted by recent studies of adults performing visuomotor tasks in natural contexts. These results suggest that the detailed physical properties of the human body convey extremely important information (Ballard, Hayhoe, Pook, & Rao 1997). Ballard et al. proposed a model of 'embodied cognition' that operates at time scales of approximately one third of a second and uses subtle orienting movements of the body during a variety of cognitive tasks as input to a computational model. At this 'embodiment' level, the constraints of the body determine the nature of cognitive operations, and the body's pointing movements are used as deictic references to bind objects in the physical environment to variables in cognitive programs of the brain. We apply the theory of embodied cognition in the context of early word learning. To do so, one needs to consider the role of embodiment from both the perspective of a speaker (language teacher) and that of a language learner. First of all, in the study of recent work (e.g. Tanenhaus, Spivey-Knowlton, Eberhard, & Sedivy 1995; Meyer, Sleiderink, & Levelt 1998; Griffin & Bock 2000; for review, see Griffin 2004), it has been shown that speech and eye movement are closely linked. Griffin & Bock (2000) demonstrated that speakers have a strong tendency to look toward objects referred to by speech, and moreover words begin roughly a second after speakers gaze at their referents. Meyer et al. (1998) found that the speakers' eye movements are tightly linked to their speech output. They found that when speakers were asked to describe a set of objects from a picture, they usually looked at each new object before mentioning it, and their gaze remained on the object until they were about to say the last word about it. Additionally, from the perspective of a language learner, Baldwin (1993) showed that infants actively gathered social information to guide their inferences about word meanings, and systematically checked the speaker's gaze to clarify his/her reference.

In our model, we attempt to show how social cues exhibited by the speaker (e.g. the mother) can play a crucial constraining role in the process of discovering words from the raw audio stream and associating them with their perceptually grounded meanings. By implementing the specific mechanisms that derive from our underlying theories in explicit computer simulations, we can not only test the plausibility of the theories but also gain insights about both the nature of the model's limitations and possible solutions to these problems.

To simulate how infants ground their semantic knowledge, our model of infant language learning needs to be embodied in the physical environment, and to sense this environment as a young child. To provide realistic inputs to the model, we attached multiple sensors to adult subjects who were asked to act as caregivers and perform some everyday activities, one of which was narrating the picture book (used in the preceding experiment) in English for a young child, thereby simulating natural infant-caregiver interactions. Those sensors included a head-mounted CCD camera to capture visual information about the physical environment, a microphone to sense acoustic signals, an eye tracker to monitor the course of the speaker's eye movements, and position

FIGURE 10.4. The computational model shares multisensory information like a human language learner. This allows the association of coincident signals in different modalities.

sensors attached to the head and hands of the caregiver. In this way, our computational model, as a simulated language learner, has access to multisensory data from the same visual environment as the caregiver, hears infant-directed speech uttered by the caregiver, and observes the body movements, such as eye and head movements, which can be used to infer what the caregiver refers to in speech. In this way, the computational model, as a simulated infant, is able to shared grounded lexical items with the teacher.

To learn words from caregivers' spoken descriptions, three fundamental problems need to be addressed: (1) object categorization to identify grounded meanings of words from non-linguistic contextual information; (2) speech segmentation and word spotting to extract the sound patterns of the individual words which might have grounded meanings; and (3) association between spoken words and their meanings. To address those problems, our model consists of the following components, as shown in Plate 8:

- Attention detection finds where and when a caregiver looks at the objects in the visual scene based on his or her gaze and head movements. The speaker's referential intentions can be directly inferred from their visual attention.
- Visual processing extracts perceptual features of the objects that the speaker is attending to at attentional points in time. Those visual features consist of color, shape, and texture properties of visual objects and are used to categorize the objects into semantic groups.
- Speech processing includes two parts. One is to convert acoustic signals into discrete phoneme representations. The other part deals with the comparison of phoneme sequences to find similar substrings and cluster those subsequences.
- Word discovery and word-meaning association is the crucial step in which information from different modalities is integrated to discover isolated spoken words from fluent speech and map them to their perceptually grounded meanings extracted from visual perception.

The following paragraphs describe these components respectively. The technical details can be found in Yu, Ballard, & Aslin (2005).

### 10.4.1 *Attention detection*

Our primary measure of attention is where and when the speaker directs gaze (via eye and head movements) to objects in the visual scene. Although there are several different types of eye movement, the two most important ones for interpreting the gaze of another person are saccades and fixations. Saccades are rapid eye movements that move the fovea to view a different portion of the

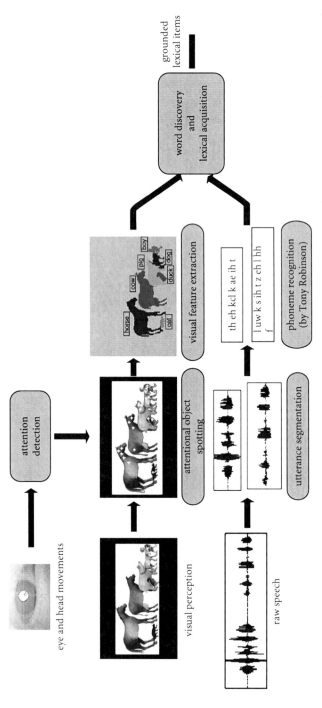

8. The overview of the system. The system first estimates subjects' focus of attention, then utilizes spatial-temporal correlations of multisensory input at attentional points in time to associate spoken words with their perceptually grounded meanings. For improved image quality and colour representation see Plate 8.

visual scene. Fixations are stable gaze positions that follow a saccade and enable information about objects in the scene to be acquired. Our overall goal, therefore, is to determine the locations and timing of fixations from a continuous data stream of eye movements. Current fixation-finding methods (Salvucci & Goldberg 2000) can be categorized into three types: velocity-based, dispersion-based, and region-based. Velocity-based methods find fixations according to the velocities between consecutive samples of eye-position data. Dispersion-based methods identify fixations as clusters of eye-position samples, under the assumption that fixation points generally occur near one another. Region-based methods identify fixation points as falling within a fixed area of interest (AOI) within the visual scene.

We developed a velocity-based method to model eye movements using a Hidden Markov Model (HMM) representation that has been widely used in speech recognition with great success (Rabiner & Juang 1989). A two-state HMM was used in our system for eye-fixation finding. One state corresponds to the saccade and the other represents the fixation. The observations of the HMM are two-dimensional vectors consisting of the magnitudes of the velocities of head rotations in three dimensions and the magnitudes of velocities of eye movements. We model the probability densities of the observations using a two-dimensional Gaussian. The parameters of the HMMs that need to be estimated consist of the observation and transition probabilities. The estimation problem concerns how to adjust the model $\lambda$ to maximize $P(O \mid \lambda)$ given an observation sequence $O$ of eye and head motions. We can initialize the model with flat probabilities, and then the forward-backward algorithm (Rabiner & Juang 1989) allows us to evaluate the probabilities. As a result of the training, the saccade state contains an observation distribution centered around high velocities, and the fixation state represents the data whose distribution is centered around low velocities. The transition probabilities for each state represent the likelihood of remaining in that state or making a transition to another state.

### 10.4.2 *Clustering visually grounded meanings*

The non-linguistic inputs of the system consist of visual data from a head-mounted camera, head positions, and gaze-in-head data. Those data provide the contexts in which spoken utterances are produced. Thus, the possible referents of spoken words that subjects utter are encoded in those contexts, and we need to extract those word meanings from raw sensory inputs. As a result, we will obtain a temporal sequence of possible referents depicted by the box labeled 'intentional context' in Plate 9. Our method firstly utilizes eye and head movements as cues to estimate the subject's focus of attention. Attention, as represented by eye fixation, is then used for spotting the target object of the subject's interest. Specifically, at every attentional point in time, we make use of eye gaze to find the attentional object from all the objects in a scene. The referential intentions are then directly inferred from attentional objects. We represent the objects by feature vectors consisting of color, shape, and texture features. For further information see Yu et al. (2005). Next, since the feature vectors extracted from visual appearances of attentional objects do not occupy a discrete space, we vector quantize them into clusters by applying a hierarchical agglomerative clustering algorithm. Finally, for each cluster we select a prototype to represent perceptual features of this cluster.

### 10.4.3 *Comparing phoneme sequences*

We describe our methods of phoneme string comparison in this subsection. Detailed descriptions of algorithms can be obtained from Ballard and Yu (2003). First, the speaker-independent phoneme recognition system is employed to convert spoken utterances into phoneme sequences. To fully simulate lexical learning, the phoneme recognizer does not encode any language model or word model. Therefore, the outputs are noisy phoneme strings that are different from phonetic transcriptions of text. The goal of phonetic string matching is to identify sequences that might be different actual strings, but have similar pronunciations. In our method, a phoneme is represented by a 15-dimensional binary vector in which every entry stands for a single articulatory feature called a distinctive feature. Those distinctive features are indispensable attributes of a phoneme that are required to differentiate one phoneme from another in English. We compute the distance between two individual phonemes as the Hamming distance. Based on this metric, a modified dynamic programming algorithm is developed to compare two phoneme strings by measuring their similarity.

### 10.4.4 *Multimodal word learning*

Plate 9 illustrates our approach to spotting words and establishing word-meaning associations, which consists of the following steps (see Yu et al. 2005 for detailed descriptions):

- Phoneme utterances are categorized into several bins based on their possibly associated meanings. For each meaning (an attentional object), we find the corresponding phoneme sequences uttered in temporal proximity, and then categorize them into the same bin labeled by that meaning.

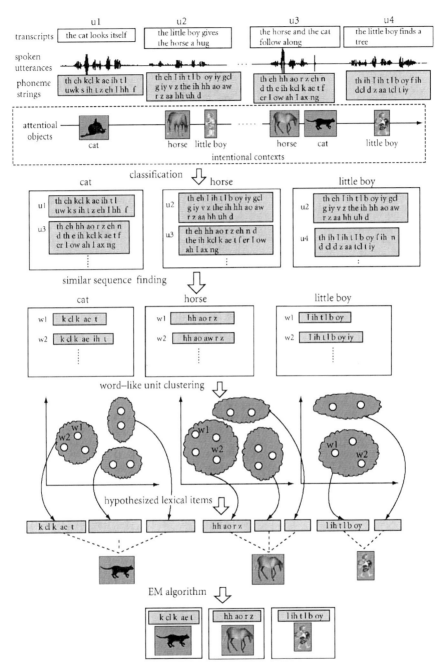

9. Overview of the method. Spoken utterances are categorized into several bins that correspond to temporally co-occurring attentional objects. Then we compare any pair of spoken utterances in each bin to find the similar subsequences that are treated as word-like units. Next, those word-like units in each bin are clustered based on the similarities of their phoneme strings. The EM-algorithm is applied to find lexical items from hypothesized word-meaning pairs. For improved image quality and colour representation see Plate 9.

- The similar substrings between any two phoneme sequences in each bin are found and treated as word-like units.
- The extracted phoneme substrings of word-like units are clustered by a hierarchical agglomerative clustering algorithm. The centroids of clusters are associated with their possible grounded meanings to build hypothesized word-meaning pairs.
- To find correct lexical items from hypothesized lexical items, the probability of each word is represented as a mixture model that consists of the conditional probabilities of each word given its possible meanings. In this way, the same Expectation Maximization (EM) algorithm described in Study 1 is employed to find the reliable associations of spoken words and their grounded meanings which maximize the likelihood function of observing the data.

### 10.4.5 *Results*

Six subjects, all native speakers of English, participated in the experiment. They were asked to narrate the picture book 'I went walking' (used in the previous experiment) in English. They were also instructed to pretend that they were telling this story to a child, so that they should keep verbal descriptions of pictures as simple and clear as possible. We collected multisensory data when they performed the task, which were used as training data for our computational model.

Table 10.1 shows the results for four measures. Semantic accuracy measures the categorization accuracy of clustering visual feature vectors of attentional objects into semantic groups. Speech segmentation accuracy measures whether the beginning and the end of phoneme strings of word-like units are word boundaries. Word-meaning association accuracy (precision) measures the percentage of successfully segmented words that are correctly associated with their meanings. Lexical spotting accuracy (recall) measures the percentage of word-meaning pairs that are spotted by the model. The mean semantic accuracy of categorizing visual objects is 80.6%, which provides a good basis for the subsequent speech segmentation and word-meaning association metrics. It is important to note that the recognition rate of the phoneme recognizer we used is 75%. This rather poor performance is because it does not encode any language model or word model. Thus, the accuracy of the speech input to the model has a ceiling of 75%. Based on this constraint, the overall accuracy of speech segmentation of 70.6% is quite good. Naturally, an improved phoneme recognizer based on a language model would improve the overall results, but the intent here is to study the developmental learning procedure without pre-trained models. The measure of word-meaning association, 88.2%, is also impressive, with most of the errors caused by a few words (e.g. 'happy' and 'look') that frequently occur in some contexts but do not have visually grounded meanings. The overall accuracy of Lexical Spotting is 73.1%, which demonstrates that by inferring speakers' referential intentions, the stable links between words and meanings can be easily spotted and established. Considering that the system processes raw sensory data, and our learning method works in an unsupervised mode without manually encoding any linguistic information, the accuracies for both speech segmentation and word meaning association are impressive.

To more directly demonstrate the role of body cues in language learning, we processed the data by another method in which the inputs of eye gaze and head movements were removed, and only audio-visual data were used for learning. Clearly, this approach reduces the amount of information available to the learner, and it forces the model to classify spoken utterances into the bins of all the objects in the scene instead of just the bins of attentional objects. In all other respects, this approach shares the same implemented components with the eye-head-cued approach. Figure 10.5 shows the comparison of these two methods. The eye-head-cued approach outperforms the audio-visual approach in both speech segmentation ($t(5) = 6.94$, $p < 0{:}0001$) and word-meaning association ($t(5) = 23.2$, $p < 0{:}0001$). The significant difference lies in the fact that there exist a multitude of co-occurring word-object pairs in natural environments that infants are situated in, and the inference of referential intentions through body movements plays a key role in discovering which co-occurrences are relevant.

TABLE 10.1 Results of word acquisition

| Subjects | Semantics (%) | Speech segmentation (%) | Word-meaning association (%) | Lexical spotting (%) |
|---|---|---|---|---|
| 1 | 80.3 | 72.6 | 91.3 | 70.3 |
| 2 | 83.6 | 73.3 | 92.6 | 73.2 |
| 3 | 79.2 | 71.9 | 86.9 | 76.5 |
| 4 | 81.6 | 69.8 | 89.2 | 72.9 |
| 5 | 82.9 | 69.6 | 86.2 | 72.6 |
| 6 | 76.6 | 66.2 | 83.1 | 72.8 |
| Average | 80.6 | 70.6 | 88.2 | 73.1 |

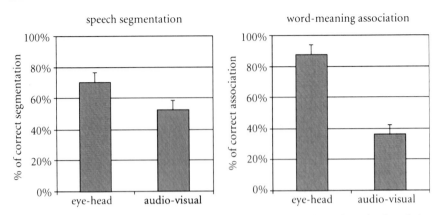

FIGURE 10.5. A comparison of performance of the eye-head-cued method and the audio-visual approach.

### 10.4.6 *Significance*

To our knowledge, this work is the first model of word learning which not only learns lexical items from raw multisensory signals to closely resemble infant language development from natural environments, but also explores the computational role of social cognitive skills in lexical acquisition. In addition, the results obtained from this comparative study are very much in line with the results obtained from human subjects, suggesting not only that our model is cognitively plausible, but also that the role of multimodal interaction can be appreciated by both human learners and by the computational model.

## 10.5  General discussion

### 10.5.1 *The role of body cues*

Children do not hear spoken utterances in isolation. They hear them in a context. Ervin-Tripp (1973) found that normal children with deaf parents, who could access English only from radio or television, did not learn any speech. Macnamara (1982) argued that it is very difficult for a child to figure out what the silent actors in interactive materials (such as a video or a TV program) are talking about. By interacting with live human speakers, who tend to talk about things that are present in a shared context with children, the child can more effectively infer what the speaker might have meant. More recently, Kuhl, Tsao, & Liu (2003) showed that American 9-month-old infants exposed to Mandarin Chinese under audio-videotape or auditory-only conditions did not show phoneme learning. Both studies indicate that learning is influenced by the presence of a live person generating body cues to attract infant attention and motivate learning. Recent experimental studies confirmed this idea, and suggested that the existence of a theory of mind could play a central role in how children learn the meanings of certain words (Baldwin 1993; Markson & Bloom 1997; Tomasello & Farrar 1986; Tomasello 2000).

In this chapter, we focused on the ability of the young language learner to infer interlocutors' referential intentions by observing their body movements, which may significantly facilitate early word learning. Clearly, this is the earliest and perhaps the lowest level of a theory of mind, and may not (at least for infants) involve any conscious knowledge that the speaker who is providing body-movement cues has explicit intentions. Nevertheless, if infants are sensitive to some of these body-movement cues, that may constrain the word-learning process sufficiently to enable it to function effectively and efficiently in early lexical development. In contrast to most other studies, our work explores the dynamic nature of body cues in language acquisition by closely resembling the natural environment of infant-caregiver interaction. In our preliminary experiment that simulated word learning using human adults, the experimenter narrated the story shown in the picture book naturally by using infant-directed speech. The adult learners were therefore presented with continuous speech and visual information as well as the dynamic movements of the speaker's gaze and head. Similarly, in our computer simulation, the computational model we built of a young language learner received continuous sensory data from multiple modalities. As we pointed out in both of these situations (adult learning and model learning), the timing of speech productions and eye movements were not perfectly aligned in these complex natural contexts. Nevertheless, the results of empirical studies showed that adult language learners exposed to a second language in the eye-head-cued condition outperformed subjects in the audio-visual condition in both word discovery (segmentation) and word-meaning tests, indicating that human subjects can utilize dynamic information encoded in the continuous body movements of the speaker to improve the learning results. How do adults take advantage of the partial, imperfect temporal synchrony between sounds and object-directed gaze? Our computational model answered this question by simulating the underlying mechanism of using body cues.

Body cues are referential in nature. In the computational model described in the previous section, a speaker's referential intentions are estimated and utilized to facilitate word learning in two ways. First, the possible referential objects defined by gaze changes in real-time provide constraints for word spotting from a continuous speech stream. Second, a difficult task of word learning is to figure out which entities specific words refer to from a multitude of co-occurrences between words and things in the world. This is accomplished in our model by utilizing speakers' intentional body movements as deictic references to establish associations between words and their visually grounded meanings. These two mechanisms not only provide a formal account of the role of body cues in word learning, but also suggest an explanation of the experimental results obtained from adult learners of a second language in our human simulation. Furthermore, the combination of human simulation and computational modeling shows conclusively that body cues serve to facilitate, and may in fact be a necessary feature of, learning the vocabulary in a new language.

### 10.5.2 *Modelling embodied word learning*

We are interested not only in what human language learners can achieve, which is demonstrated in Experiment 2, but also in how they do so. Theoretical simulation studies provide unique opportunities to explore the mechanistic nature of early word learning, to provide a quantitative computational account of the behavioral profile of language learners, and to test hypotheses quickly (i.e. without requiring the collection of new data). Therefore, computational investigations of language acquisition have recently received considerable attention. Among others, MacWhinney (1989) applied the competition theory to build an associative network that was configured to learn which word among all possible candidates refers to a particular object. Plunkett, Sinha, Moller, & Strandsby (1992) built a connectionist model of word learning in which a process termed 'autoassociation' maps preprocessed images with linguistic labels. The linguistic behavior of the network exhibited non-linear vocabulary growth (vocabulary spurt) that was similar to the pattern observed in young children. Siskind (1996) developed a mathematical model based on cross-situational learning and the principle of contrast, which learns word-meaning associations when presented with paired sequences of pre-segmented tokens and semantic representations. Regier's work (1996) focused on grounding lexical items that describe spatial relations in visual perception. Bailey (1997) proposed a computational model that can not only learn to produce verb labels for actions but also carry out actions specified by verbs that it has learned. Tenenbaum & Xu (2000) developed a computational model based on Bayesian inference which can infer meanings from one or a few examples without encoding the constraint of mutual exclusion.

Computational models of development and cognition have changed radically in recent years. Many cognitive scientists have recognized that models which incorporate constraints from embodiment—i.e. how mental and behavioral development depends on complex interactions among brain, body, and environment (Clark 1997)—are more successful than models which ignore these factors. Language represents perhaps the most sophisticated cognitive system acquired by human learners, and it clearly involves complex interactions between a child's innate capacities and the social, cognitive, and linguistic information provided by the environment (Gleitman & Newport 1995). The model outlined in the present study focuses on the initial stages of language acquisition using the embodied cognition perspective: how are words extracted from fluent speech and attached to meanings? Most existing models of language acquisition have been evaluated by artificially derived data of speech and semantics (Brent & Cartwright 1996; Siskind 1996; Regier 1996; Cohen, Oates, Adams, & Beal 2001—but see also Roy & Pentland 2002). In those models, speech is represented by text or phonetic transcriptions and word meanings are usually encoded as symbols or data structures. In contrast, our model proved successful by taking advantage of recent advances in machine learning, speech processing, and computer vision, and by suggesting that modeling word learning at the sensory level is not impossible, and that embodiment has some advantages over symbolic simulations by closely resembling the natural environment in which infants develop. In both empirical and computational studies, we use storybook reading—a natural interaction between children and caregivers—to simulate the word learning in everyday life. Multisensory data (materials used by the model) are real and natural. To our knowledge, in the literature of language acquisition modeling, this experimental setup is the closest to the natural environment of early word learning that has been achieved.

Our model emphasizes the importance of embodied learning for two main reasons. First, the motivation behind this work is that language is grounded in sensorimotor experiences with the physical world. Thus, a fundamental aspect of language acquisition is that the learner can rely on associations between the movements of the body and the context in which words are spoken (Lakoff & Johnson 1980a). Second, because infants learn words by sensing the environment with their perceptual systems, they need to cope with several practical problems, such as the variability of spoken words in different contexts and by different talkers. To closely simulate infant vocabulary development, therefore, a computational model must have the ability to remove noise from raw signals and to extract durable and generalizable representations instead of simplifying the problem by using consistent symbolic representations (e.g. text or phonetic transcriptions). Furthermore, our computational model addresses the problem of speech segmentation, meaning identification and word-meaning mapping in a general framework. It shows the possible underlying mechanism by which linguistic processing, perceptual learning, and social communication interact with each other in early word learning.

## 10.6 Conclusions

All three of our studies show quantitatively how body cues that signal intention can aid infant language learning. Such intentional body movements with accompanying visual information provide a natural learning environment for infants to facilitate linguistic processing. From a computational perspective, this work is the first model that explicitly includes social cognitive skills in language learning, such as inferring the mother's referential intention from her body movements. The central ideas of our model are to identify the sound patterns of individual words from continuous speech using non-linguistic contextual information and employ body movements as deictic references to build grounded lexical items. By exploiting the constraints of social interaction and visual perception, probabilistic algorithms such as expectation maximization have the power to extract appropriate word-semantics associations even in the highly ambiguous situations that the infant normally encounters. Equally important is that the model suggests a framework for understanding the vocabulary explosion that begins at age 2. Besides providing a relatively limited number of the most probable lexical items, the EM model also generates a large amount of word-meaning pairs with uncertainty. This indicates that infants can potentially accumulate valuable information about many word-semantics associations long before these associations are unique. The rapid vocabulary expansion may be a product of this parallel accumulation process.

# 11

# Talk About Motion: The Semantic Representation of Verbs by Motion Dynamics

ERIN N. CANNON AND PAUL R. COHEN

## 11.1 Introduction

Humans are perceivers and cognizers in an ever-changing dynamic world. Every moment is unique and different. How are we able to make sense of our experiences, label them with words, and speak in a way that is meaningful to others? If we labeled every situation uniquely, then the number of words in the human vocabulary would be infinite, making the utterance of a word not only uncommunicative, but essentially meaningless. Therefore, for purposes of effective communication with others, we cannot view every situation as unique. There must be commonalities between situations that call for the same words to be uttered in slightly different situations, and conversely, for words with slightly different meanings to be appropriate in overlapping situations. A situation, which we will call **s**, must be an abstraction of some kind. This chapter will explore the role of motion dynamics in making these abstractions available through perception, based on the assumption that dynamic real world movement is a reliable cue providing meaning about the world. In the case of action words (i.e. verbs), we assert that the dynamical movement of objects through space provides the semantics for the words we choose to utter.

This chapter focuses on abstractions from patterns of movement that give rise to the utterance of verbs. That is, we consider the possibility that situations, **s**, are representations containing abstractions of movement patterns. We begin by putting forth a theory of word meaning suggested by Oates (2001), which is based on the ideas of pattern extraction, and a new way for cognitive scientists to view the questions of semantic language learning. We then review literature that spans the fields of social, cognitive, and linguistic development, which demonstrates that humans are remarkably sensitive to patterns of movement in space. We survey what is known about neonatal abilities to discriminate patterns of movement. Then we look at how different languages may influence movement patterns attended to. Finally, we present in detail one account of **s**, Cohen's Maps for Verbs framework, and discuss empirical evidence for it.

## 11.2 Word meaning

The choice of words is conditional: One is more likely to say 'dog' than 'Thursday' when a dog is present, even if 'Thursday' has a higher unconditional probability of being uttered. Informally, the choice of words is conditioned on the situation—a dog is present, or someone asks what day it is. It is difficult to think of situations that *determine* particular utterances. In general, a word has a *probability* of being uttered given the situation, which includes the words that have been uttered. Following Oates (2001) we define the *meaning* of a word as this propensity to be uttered in a situation. What does 'Thursday' mean in a given situation? It means that something in the situation makes 'Thursday' a likely word to be uttered. In general, the probability of uttering word **w** in situation **s**, $Pr(utter(\mathbf{w}) \mid \mathbf{s})$, is not the same as the probability that **s** is true given that **w** has been uttered—$Pr(\mathbf{s} \mid utter(\mathbf{w}))$—but these probabilities are proportional to one another, as any intuitive account of word meaning requires.[1]

The general form of this theory of word meaning might be right, but lacks three specifics. First, the probability that a word will be uttered depends not only on the situation but also on the speaker. What we really need is $Pr(utter(\mathbf{p},\mathbf{w}) \mid \mathbf{s})$ for every person **p**. Of course, we cannot have this information, so we must approximate it. Oates (2001) describes how to make the approximation. Second, this simple theory of word meanings does not explain how compositions of words (e.g. sentences) have meanings. This chapter says nothing about syntax and the composition of words into sentences. Third, the

---

[1] From Bayes' theorem we have

$$P(utter(w) \mid s) = P(s \mid utter(w)) * P(utter(w)) / P(s)$$
$$P(s \mid utter(w)) = P(utter(w) \mid s) * P(s) / P(utter(w))$$

These expressions correspond to language generation and understanding, respectively. The first governs the probability that one will say a word in a given situation, the second is used to infer which situation holds given that a word is spoken. These conditional probabilities are clearly proportional, each is a scaled version of the other, where the scaling is by a ratio of two prior probabilities, the unconditional probability of the situation and the unconditional probability of uttering the word. For a given $P(utter(w) \mid s)$, the probability of s given **w** is proportional to the unconditional probability of s and inversely proportional to the probability of uttering **w**. This latter condition is another way of saying that the word **w** carries information about the situation s: The less likely one is to utter **w**, the more likely it makes s given **w**.

theory does not specify the elements of situations that go into **s**, the propositions on which word choices are conditioned. However, by bridging the fields of cognitive development and language acquisition, we can hypothesize and test potential candidates for **s**. This is the goal we set forth in this chapter, and for guiding future research.

We do not suppose that patterns of movement are the *only* elements of situations **s** on which word choices are conditioned. Presumably **s** contains other physical observables such as the number, shape, and classes of objects. Complicating the story, **s** might also contain *unobservable* elements, particularly attributions of beliefs and goals. Suppose one observes George walking down the street a few yards behind Fred. The word 'follow' is ambiguous in this context. It might mean only that George is walking behind Fred, or it might mean that George intends to walk behind Fred and go wherever Fred goes. Let us assume that nothing in Fred's or George's observable behavior indicates that George is following Fred in the second sense of the word, and yet a speaker, observing the scene, decides to use this sense of 'following'; indeed, the speaker might even say 'tailing' or 'stalking', or some other word that indicates George intends to stay close to Fred as he walks along. If the choice of words is conditioned on a representation of the situation, **s**, then **s** must contain an attribution of George's intention to remain close behind Fred. Of course, this attribution might be wrong (e.g. a false belief), but it is an element of **s**, and therefore contributes to the word choice uttered.

Intentional words complicate an otherwise straightforward theory of the acquisition of word meanings. If word choices are conditioned on *observable* aspects of the situation, **s**, then a child could learn word meanings by associating words with situations, that is, by learning conditional probabilities $Pr(utter(\mathbf{w})|\mathbf{s})$. However, if word choices are conditioned on *unobservable* aspects of situations, then associative learning is more difficult. Suppose a child observes a dog running after a squirrel while her mother says, 'The dog is chasing the squirrel.' One can see how the child might learn to associate 'chasing' with the observable, physical aspects of the scene—both animals are running, when the squirrel changes direction the dog does, too—but how can the child learn that 'chasing' implies something about the intentional states of both the dog and the squirrel, when these states are not observable? Presumably, at some point in the child's development, she is able to supply these unobservable elements, herself. She *imagines* the intentional states of the animals and associates these states with the word 'chasing'. The problem with this theory is that it is difficult to prove, because it asserts that the child conditions her word choices on intentional states she imagines, and we cannot observe what she imagines. More concretely, we cannot be sure that, to a young child, 'chasing' does *not* mean only the physical aspects of chasing, nor can we easily discover when, in the child's development, the meaning is extended to include intentional aspects of the situation.

In fact, it is difficult to interpret some of the literature that seems relevant to our claim that word choices might be conditioned on patterns of movement. The general problem has this schematic form: Infants or older children are shown to discriminate patterns of movement, say $P_1$ and $P_2$, which adults label with intentional terms, such as 'avoid' or 'pursue'. Presented with $P_1$ and $P_2$, what discrimination is the infant, child, or adult *really* making? The adult might be comparing the raw movement data, $P_1$ vs. $P_2$, or she might be comparing her intentional interpretations of $P_1$ and $P_2$, or both. In one case we say that the adult discriminates the dynamics of the displays, in another we say that the adult discriminates 'avoid' and 'pursue'. We do not know which is true, and both might be. The same goes for the infant and the child: We cannot say *when* or even *whether* intentional attributions inform discriminations of displays, particularly when displays might be discriminated based on (even subtle) differences in dynamical motion. We should not assume that, because adults make intentional attributions to displays, the child's ability to discriminate entails discriminating intentional states.

## 11.3  Review of the literature

We begin with Heider & Simmel's (1944) classic demonstration that patterns of movement evoke rich linguistic descriptions. Evocation is a phenomenon, not an explanation. We cannot say *why* subjects find so much to say about Heider & Simmel's displays. However, the only information-carrying aspect of the display is the relative movement of a few shapes. The lengthy and imaginative stories about the displays must be cued somehow by these movements. Next, we review work based on point-light displays, which shows that humans can reliably extract movement information in the absence of shape cues. Having established humans' sensitivity to patterns of movement, we build a case that these patterns support *semantic* distinctions, including differences in word meanings. Infants can discriminate patterns of movement generated by different classes of things, and young children appear to discriminate causal from non-causal movement in launching events. The patterns available to neonates are candidates for elements of **s**, the situation descriptions on which probabilities of uttering words are conditioned. This literature gets us ready for linguistic theories in which word meanings are grounded in physical dynamics. We review these theories, including developmental arguments. We then discuss the ways in which a scene is parsed into meaningful motion-based components, which will inform **s**. In conclusion, further candidates for the semantic core are suggested in P. R. Cohen's Maps for Verbs framework.

### 11.3.1 *Patterns of movement evoke intentional descriptions*

In Heider & Simmel's (1944) classic study, adults were shown a film clip of three shapes in motion. The adult participants created elaborate storylines describing the interactions, even though the only information in the stimuli was object shape and motion. Human-like characteristics were easily attributed to the triangles and circles, including intentional states. Moreover, the critical phenomenon discovered in this study is that the attributions given to each shape were highly similar across participants. All reports included common event features: a fight scene, a chase scene, and a scene in which one object became trapped in the house and tried to escape. Thus, not only did these simple motion patterns elicit detailed anthropomorphized descriptions and storylines, but the actual verbal reports were similar. Although Heider & Simmel did not test for similarities between particular utterances, their findings suggest that movement patterns may predict *which* intentional descriptions are attributed to them.

If adults have tendency to extract intentional attributes from patterns of movement or events, then so might children. Berry & Springer (1993) tested 3–5-year-olds to investigate the influence of motion dynamics on anthropomorphic attributions. Four groups of children were tested systematically. One group received the original Heider & Simmel movie, another received the movie with the object shapes obscured, preserving only the motions; the third group received static displays taken from the movie, with shapes and figure information preserved; and the last group received static displays where both shape and motion were obscured. The experimenters obscured the shapes of objects to rule out the possibility that object shape or size contributed to the characteristics attributed to the objects. While watching the film, children were asked, 'What do you see?' Like adults, children attributed intentions to the objects in the movies, and were about five times more likely to use anthropomorphic language, including intentional attributions, than children who were shown static displays. Shape did not seem to be a relevant factor in the intentional attributions. Clearly, then, by the age of 3, motion is a sufficient cue to determine word choices whose meanings convey intention.

Two factors make these findings quite compelling. First, an understanding of intentionality is a prerequisite to children's theory of mind (TOM; e.g. Leslie 1984), yet three-year-olds have difficulty understanding that other people's intentions may vary from their own (particularly about beliefs, it may be less difficult for desires; see Bartsch & Wellman 1995; or Flavell 1999, for review of

TOM literature). It is curious, then, that young children so adamantly ascribed intentional states (as indicated by their word choice) to the moving shapes in the Heider and Simmel movie. Berry & Springer did find a trend toward increasingly anthropomorphic descriptions with age, but it did not reach significance. It might be fair to say this that some portion of the anthropomorphic descriptions, then, did come from 3-year-olds. Second, the task was not forced-choice: children gave open-ended descriptions of the films they watched. These children were young language learners, with a far more limited vocabulary than adults. Yet even by the age of 3, their choice of words to describe the scene was remarkably adult-like with respect to intentional attributions. This suggests that the children were no less able than adults to extract the motion patterns that elicited their word choices.

More compelling is that even preverbal infants show an ability to extract intentional information from movement patterns (e.g. Golinkoff & Kerr 1978; Legerstee, Barna, & DiAdamo 2000; Leslie 1984; Spelke, Phillips, & Woodward 1995; Woodward 1998). Intentional attributes have been suggested in habituation and violation-of-expectation paradigms focused on the understandings of goal-directed actions and concepts of agency. Both goal-directedness and a concept of agency implies that intentionality is involved in a scene. One difficulty, however, is the confound of infants' familiarity with human actions. Humans are inherently agents, thus intentional beings, and are also often the subjects in these experiments. However, non-human and inanimate objects have been successfully utilized to serve as 'agents' in motion events also (e.g. Cohen, Rundell, Spellman, & Cashon 1999; Cohen & Oakes 1993; Gergely, Nadasdy, Csibra, & Biro 1995). In some cases, infants *may* perceive inanimate objects as intentional, based solely on particular motion characteristics such as self-propulsion and trajectory (Baron-Cohen 1994; Premack 1990), or by moving along a trajectory through space in a 'rational' manner (Gergely et al. 1995; Csibra, Gergely, Biro, Koos, & Brockbank 1999). As touched upon in the introduction, we cannot be sure that the discrimination of intentional states is the same in early childhood and infancy as it is in adulthood. The ability to use motion dynamics for discriminating goal directedness and agency early in life, however, is suggestive that attributions of intentionality begin prior to the first words being uttered. It could be that some unknown is present in the motion dynamics, or that something draws the infant to attend to particulars of the motion specifying intentionality. Children may learn to attach intention-loaded words to these motions, perhaps even before they fully understand the implications of that particular word. As vocabulary increases, so does the child's understanding of intentionality—which probably develops from a motion-based understanding—to more of a psychologically-based and adult-like understanding. In experiments such as Heider & Simmel's, perhaps the motion-based elements in **s** are substantial enough to elicit the intentional words that were associated with them earliest in development.

### 11.3.2 *Sensitivity to patterns of movement*

The work of Johansson (1973) proposed that the visual system parsed biomechanical movement presented in point-light displays into two separate types of motion: *common* motion, from which the trajectory of the group of lights relative to the observer is perceived, and *relative* motion, the invariant relations between these lights, from which structure, or figure, is perceived. Indeed, using similar point-light displays, Bertenthal, Proffitt, & Cutting (1984) found that infants as young as 3 months discriminated biological motion, specifically the relative motion patterns of human walkers. In a habituation (with partial lag) experiment, infants were able to discriminate upright human walkers from inverted human walkers, but they could not make this discrimination when tested with static light displays. The infants evidently extracted figural coherence from information in the moving displays. In a second experiment, absolute motion was held constant, and thus the only motion information available was the relative motion from the light points. In this experiment, infants were able to discriminate the real walkers from anomalous, scrambled points of light. Moreover, infants were not using *absolute* motion cues in the detection of biomechanical motion. These findings suggest that perception of patterns of relative motion is functioning early in life. It is not unreasonable to assume that this information is extracted and utilized to inform and create semantic representations about the world as the child experiences it.

Additionally, Bertenthal (1993) suggested there might be several other processing constraints responsible for biomechanical motion perception that are available to the perceptual system early on. For instance, a slight spatial discrimination seems not to affect infants' discriminations of biological motion (disruptions of local rigidity), but temporal disruptions in the movements of individual points of light do in fact disrupt this perception. Bertenthal & Pinto (1994) found similar results when testing adults; temporal disruptions made to the individual points of light impaired the perception of biological motion, more so than spatial disruptions, supporting the idea that motion is extremely important in information extraction. In addition, the influence of stimulus familiarity also constrains biomechanical motion perception (Bertenthal 1993). When tested with non-human biological motion, in this case spiders, 3-month-olds discriminated inverted displays from upright ones but 5-month-olds did not. Bertenthal attributes this discrepancy to a shift in perceptual processing by 5 months to a level based on 'perceived meaning' (p. 209).

Sensitivity to specific patterns of motion containing meaning is not exclusive, however, to biological motion. As discussed earlier, the non-biological pattern of motions presented by Heider & Simmel (1944) elicited responses *as if* the objects themselves were 'biological'. Guylai (2000) found that manipulating the kinetic patterns of movement between objects in a 2D movie display influenced the attributed meanings (based on specific questions asked to participants about the event) more so than changes to object hue, size, shape, or luminance. Other perceptual cues did not change the overall impression. It appears that kinetic patterns amongst objects (or points) influence how we perceive the content or meanings of events.

### 11.3.3 *Semantic core and patterns of movement*

In the introduction to this chapter we suggested that word choices are conditioned in part on representations of the current scene, which we denoted **s**. Representations are constructed from elements, and we are particularly interested in the most *primitive* elements, the ones infants might have or learn. Several cognitive scientists think these elements may be learned through interaction with the physical world (Barsalou 1999; Johnson 1987; Mandler 1992; 2000). In the following sections we will survey some candidates for these primitive representational elements, which we call the semantic core, and then show how these might serve to specify the meanings of words (P. Cohen et al. 2002; Oates 2001). We are particularly interested in those primitive semantic distinctions that can be grounded in patterns of movement.

### 11.3.4 *Motion and causality*

Michotte (1963) suggested that the perception of causality could be manipulated. His simple animations of two squares interacting suggested that causality is perceived directly, without cognitive interpretation. Of particular interest here is the *launching event*. Perceived as a whole-body interaction, a launching event is one in which object A moves toward a static object B, stops at the point of contact, and then object B appears to be set into motion as a result. Adults report perceiving this sort of event as causal, in that object A caused the movement in object B. When Michotte manipulated temporal and/or relative velocity patterns, interactions were perceived as qualitatively different. For example, if object B began to move within 70 msec. of contact, its movement was perceived as causally related to the interaction with object A. If object B moved after 160 msec., then its movement and A's movement were perceived as disconnected, not causally related. Similarly, manipulating the gap between the two objects just prior to the movement of the second one, or their velocities, affected whether the interactions were perceived as causal or separate autonomous movements. Thus highly specific spatio-temporal features of interactions affect whether events are perceived as causal or not.

The ability to detect spatio-temporal features of interactions is present early in life (Leslie 1982; 1984). Young infants tested in a habituation paradigm were shown Michottian launching events, with manipulations of delays at contact and spatial gaps. Leslie (1984; 1988) suggested the ability to detect the internal structure of a launching event was present by 6 months of age. Six-and-a-half-month-olds habituated to a launching event then dishabituated to events involving a spatial gap plus a temporal delay. However, infants habituated to a delayed launch did not dishabituate to scenes involving a spatial gap, and vice versa (Leslie 1984). These infants showed sensitivity to specific disruptions in spatio-temporal continuity. Leslie & Keeble (1987) supported this notion by reversing the direct and delayed launching events. Six-month-olds were habituated to a film clip of a red square directly launching a green square. Then the clip was played backwards. The reasoning goes that a causal event (the direct launch) involves an agent (the causer of an action) and a recipient of that action. Reversal of the causal event involves a reversal also, of the mechanical roles. A second group of infants was habituated to a delayed launch, then tested on the film played backwards. If the event was not perceived as causal, then there should be no change in role reversal either. The hypothesis was confirmed; infants dishabituated in the direct launching condition, but not to the action reversal in the delayed launching. Leslie & Keeble (1987) concluded that infants discriminated on the basis of causal relations.

Whereas Leslie wants to argue from a modularity perspective that causality is a primitive concept (e.g. Leslie 1994), the work of L. Cohen and colleagues (e.g. L. Cohen & Oakes 1993; L. Cohen & Amsel 1998; Oakes 1994) suggests that the perception of causality is actually developmental and is built up from simpler percepts. In terms we introduced earlier, the semantic core would include these simpler percepts and the launching event itself would be what we have called **s**, the situation. Here we will briefly review the evidence that infants perceive components of a launching event.

Cohen & Amsel (1998) investigated the development of causal perception for infants slightly younger than those used in Leslie's (1984) experiment. They tested for changes in habituation from direct launching events to both types of non-causal event—those with a temporal delay and those with a spatial gap. Note that these discriminations are more finely tuned than Leslie's non-causal events involving both spatial gaps and delays. They found that 4-month-olds did not dishabituate to non-causal events, but showed a general preference for looking to causal events. By five-and-a-half months, infants dishabituated to change in any feature, causal or non-causal. By six and a quarter months, infants dishabituated on the basis of causality only. Oakes (1994) also found that by 7 months, infants discriminated on the basis of causality only, and not as a response to changes in independent features.

However, the ability at 6 and 7 months of age to discriminate events on the basis of causality is not particularly strong. At this age, it is fairly situation-specific. For example, Oakes & L. Cohen (1990) tested the perception of causal events using complex stimuli, more like objects in the real world (as opposed to animated squares and such). Six-month-olds did not dishabituate on the basis of causality in this case, but 10-month-olds did. Furthermore, Oakes (1994) found that 7-month-olds did not discriminate on the basis of causality when the paths or trajectories of the objects in the event varied. By 10 months, infants were not bothered by changes in path, but did discriminate on basis of causality. But even at this age, L. Cohen & Oakes (1993) argue that causality is still somewhat tied in with object perception. For example, 10-month-olds tended to respond differentially to changes in identity of the objects before generalizing the event in terms of causality.

Taken together, the literature on perception of physical causality suggests that, by the end of the first year, causal perception is nearly adult-like. Furthermore, it has a developmental trend: There is an initial preference for responding to causal events, perhaps making the infant pay attention to them. Then, early on, there is detection of subcomponents of the event. This is the time at which infants learn which features of scenarios make up causal versus non-causal events. These spatial and temporal features are perhaps components of the semantic core, as each component conveys meaning. Once the child can assemble them into representations of situations, **s**, responses tend to be no longer based on the individual features themselves, but rather on **s**. However, instances of **s** are initially situation-specific, then abstracted, as other developing elements of **s** (such as object and agency concepts, which happen to also draw upon spatiotemporal components of the semantic core) are also refined.

### 11.3.5 *Motion and classification*

As described, a situation **s** can be parsed into elements of the semantic core. We have seen that elements of a situation are the basis for judgements of physical causality. Now we consider elements that might account for both object and action *classes*.

The categorical distinctions children (and perhaps infants) make are based on the different types of motion pattern that become associated with a particular class. We are certainly not the first to make this claim (see Lakoff 1987; Mandler 1992; 2000; Rakison & Poulin-Dubois 2001). A central example is the animate/inanimate distinction. Mandler (1992) proposed a semantic core composed of primordial image schemas to account for the animate-inanimate class. These schemas are based on motion properties, such as motion trajectory, in relation to ground and other objects, and self-propulsion. Rakison & Poulin-Dubois (2001) provide a different, perceptually based associationist explanation of how the distinction develops, which includes the properties Mandler asserts, in addition to properties such as goal-directedness and agency. Others have also considered animacy as derived from object motion in the absence of physical or mechanical causality (Leslie 1994; Premack 1990). For example, Premack (1990) suggested that if an object's change of movement is self-propelled, and not due to the movement of any other objects, then it is perceived as intentional. If both objects are self-propelled, then they might be perceived as one object being directed by the goal to affect the other object.

One issue is whether the animate/inanimate distinction is purely perceptual or whether it is knowledge-based. Perceptual categorization is based only on physical features of objects, and requires no knowledge of object function, or of what the object *is*. Mandler (1992; 2000) proposed that the behaviors demonstrated by young children are guided by conceptual knowledge about objects in the physical world, an understanding of what they are. Mandler suggested that conceptual knowledge is produced by perceptual redescriptions based on the primordial image schemas.

Much of the animate/inanimate distinction research has been based on discrimination between two domains: animals and vehicles. Objects in these domains can be perceptually similar (e.g. birds and airplanes) or perceptually dissimilar (horses and motorcycles). However, the motion patterns of the animal domain are different from the motion patterns of the vehicle domain. For instance, the pendular motion of animals is quite different from the rotary motion of vehicles. While much research favoring Mandler's conceptual knowledge has involved the extended imitation paradigm (e.g. Mandler & McDonough 1996), and has found children to make distinctions toward the end of the first year, it is unclear that motion cues are the basis. The objects tested are not actually moving in the experiment. It is quite possible that the distinction is made early, but the nature of the paradigm makes this difficult to test. The image schemas, however, are not necessarily 'knowledge-rich' in the sense that this paradigm tests for. Image schemas are dynamical—about movement and change. They are the semantic primitives that distinguish situations, s, thus organizing the knowledge acquired in these learning situations. An alternative approach, the use of point-light displays, has been an effective means of determining whether motion cues alone are a sufficient basis for the classification of animals and vehicles.

Arterberry & Bornstein (2001) tested 3-month-old infants in a multiple-exemplar habituation paradigm to search for evidence of a possible categorical distinction between animals and vehicles made at this early age. Furthermore, they tested whether this distinction was based primarily on the dynamic motion features inherent in these domains (by using point-light displays of animals and vehicles in motion) or on static featural information (pictures of animals and vehicles). The infants in both conditions dishabituated to novel categories, suggesting that they are making the animal/vehicle distinction early. Because they dishabituated in both the static and dynamic conditions, an animate/inanimate distinction could not be claimed. The figural features in the static pictures, such as legs versus wheels, could not be ruled as a basis for classification in this study.

In a similar task, Arterberry & Bornstein (2002) tested 6- and 9-month-olds in the same paradigm. Six-month-olds again showed the ability to categorize animals and vehicles based on either static or dynamic features. However, only 9-month-olds showed transfer between these display modalities. Nine-month-olds who were habituated on dynamic motion features were then able to transfer this knowledge to static displays at test. However, if the 9-month-olds were habituated to static displays of animals or vehicles, they did not transfer the categorical distinction when tested with dynamic motion displays of those animals or vehicles. This suggests that (1) there is a developmental aspect to this categorization, (2) dynamic motion conveys more transferable information than the figural features available in static displays, and (3) the transference of discriminations based on dynamic features over to static displays suggests that the children somehow 'connect' the figural information in the static displays with the dynamic information. The ability fits nicely into our theory that dynamic features represented in the semantic core are easily transferred into new instances of **s**.

### 11.3.6 *Linguistic research and cognitive semantics*

Thus far, we have discussed possible elements of **s**, the situation description which is constructed from elements of a semantic core. We have focused on psychological evidence that the semantic core contains abstractions of patterns of movement. We have not discussed linguistic issues, particularly our characterization of word meaning as the conditional distribution of a word given situations **s**. In this section we review evidence that patterns of motion influence the choice of words, that is, the proposition that **s** contains representations of patterns of motion.

Talmy coined the term 'force dynamics' (1975; 1988; 2000) to denote a semantic category that covers a full range of relations that any object or entity can have with respect to some force imposed on it. Force dynamics pertains to motion events involving two objects that are broken into linguistic primitives of causation, but further allows for other concepts such as letting or resisting. Talmy's framework includes such concepts as the exertion of force, amount of resistance, obstructing force, and overcoming resistance. Talmy (1975) claimed that there are universal structures in all languages, reflecting motion situations in which one object is moving or located with respect to another object. The motion situation is universally encoded by the following four components: (1) Figure, (2) Ground, (3) Path, and (4) Motion. Of particular interest here to the issue of verb usage are Path and Motion. Figure and Ground are typically expressed as nouns. Talmy (1988; 2000) described *verb-framed* languages as those that conflate path with motion, meaning that verbs usually express or encode path. Spanish is an exemplar. In contrast, *satellite-framed* languages tend to conflate manner with motion, as in English. Work by Naigles, Eisenberg, Kako, Highter, & McGraw (1998) found these typological differences in verb usage demonstrated by English and Spanish adult speakers when presented dynamic motion events.

### 11.3.7 *Developmental linguistics*

The story so far is that motion is an important component of word meanings. It is one of the elements of situations **s** that influence the probabilities of uttering or hearing particular words. On this account, learning word meanings is just learning conditional probability distributions Pr(utter(**w**) | **s**). However, this account cannot be complete, because it does not explain why children in different language communities do not learn the particular kinds of word in roughly the same order. Let us assume that American (native English-speaking) and Korean (native Korean-speaking) children have roughly the same experiences: Both live in a world of surfaces, objects, movements, physical influences and control, animate and inanimate motion, and so on. Thus, the situations **s** to which the children are exposed are the same. The words to which they are exposed are different, but the *kinds* of word—nouns, verbs, and so on—are not. Let us modify our account of word meaning a little to include word classes. The meaning of a particular verb class, say, is just the probability distribution over uttering a verb in that class given the situation: Pr(utter(verb class) | **s**). If lexical learning is no more than learning these conditional distributions, then Korean and American children should learn identical distributions for identical word classes. After all, the children are exposed to the same situations, **s**, so if both learn a particular verb class **v**, they should learn the same conditional distributions Pr(utter(**w** in **v**) | **s**). However, American and Korean children do not map elements of **s** to word classes in the same way, nor do they learn instances of word classes in the same order.

Choi & Bowerman (1991) found evidence that nouns are not always acquired before verbs, as previously thought (e.g. Gentner 1978; 1982). Diary accounts of English and Korean learning children were examined, and differences in verb acquisition tended to reflect the language they learned. The data suggested an interaction between young children's linguistic input (i.e. the language they are learning) and cognitive development. Korean is a verb-framed language, in which Path is typically expressed in the main verb and Manner expressed separately. English, a noun-based, satellite-framed language, expresses Manner in the main verb and Path separately. Choi & Bowerman (1991) concluded that an initial sensitivity to the semantic structures of a language is responsible for differences in language acquisition. A simple mapping of learned words to semantic elements (e.g. Slobin 1973) cannot fully account for the meanings of children's spatial words (in this study) being language-specific. Learning the lexicon might in fact mean learning conditional distributions Pr(utter(**w** in **v**) | **s**), but we still must explain how a Korean word class is conditioned on the element of **s** we call Path while the same word class in English is conditioned on an element of **s** called Manner.

The work of Tardif and colleagues (Tardif 1996; Tardif, Shatz, & Naigles 1997) suggested that noun/verb differences in language acquisition between English and Mandarin learners could be explained by looking at the linguistic input (e.g. proportion of nouns and verbs spoken) from the caregiver. Mandarin-speaking caregivers tended to produce more verbs than nouns when speaking to their children. In turn, this bias was reflected in children's vocabulary development. Work by Hoff (2003) found that environmental input factors, other than language *type*, should also be considered. Within the English-speaking population, her work has found influences of maternal speech (i.e. linguistic input) on vocabulary development as a function of socioeconomic status (SES). Specifically, children with higher SES had vocabularies that were larger and faster-growing than lower SES children. Differences were present by the age of 2, and were linked to the frequency and length of mothers' utterances to (and with) the child. Tomasello (1992; 1995) further emphasized the importance of the social context in verb learning, pointing out that children best learn from the observations of other people's actions and through their social interactions with others.

In addition to vocabulary development, Gopnik & Choi (1995) have shown a direct effect of language's influence on development of cognitive structures. Korean mothers tend to use more relational terms and action verbs when talking to their children, whereas English-speaking mothers tend to initially label objects most often with their young children. They noted that Korean children have a 'verb spurt' analogous to the noun spurt in learners of the English language. Consequently, these differences in vocabulary were reflected in children's cognitive development. Korean children showed means-ends skills earlier than English-learning children, but the English-learning children showed more advanced skills in an object categorization task.

In sum, the idea that lexical acquisition involves learning conditional probabilities Pr(utter(w) | s) is not necessarily wrong, but it does not explain how individual languages select particular elements of a situation s to serve as the features that condition word probabilities. We have already seen that Manner is a conditioning element of s for English verbs whereas Path is a conditioning element of s for Korean verbs. Nothing in our theory of word meanings yet explains this difference.

### 11.3.8 *Parsing the scene*

The challenge is to explain how elements of the semantic core—the most primitive distinctions—are collected into situation descriptions s, and to explain why these elements are bundled in different ways in different languages. We assume that all humans have access to the same elements of the semantic core; for example, American and Korean children are equally able to detect the Path or Manner of motion. It might be that the apparent differences in how English and Korean bundle elements of the semantic core are all explained by simple associative learning. This is how it might work: An English-speaking child and a Korean child are both observing the same situation, and both hear verbs with essentially the same meanings, but the best account of the verb meaning for the English speaker is obtained by conditioning the probability of the verb on elements of the scene called Manner, while the best account for the Korean child is had by conditioning the probability on Path. In this context 'best account' means 'maximizes discriminability'. Put differently, the English-speaking child will be more able to discriminate verbs by attending to Manner, while the Korean child will prefer to attend to Path. If this happens often enough, then Manner will become an important element of s for English speakers and Path will serve the same purpose for Koreans.

If this account is correct, then it will appear as though the child has rules for parsing a scene into situation descriptions s, and these rules are related to the child's native language. The rules are illusory, however. Students of each language simply search for those elements of the semantic core that best explain why words are used in particular scenes. Recent work suggests that certain motion cues and intention-based actions predict where a scene may be parsed

(Baldwin, Baird, Saylor, & Clark 2001; Zacks 2004), but says nothing about the role of language.

Evidence that these linguistic elements are accessed in motion events has recently been studied in young children and infants. Golinkoff, Chung, Hirsh-Pasek, Liu, Bertenthal, Brand, Maguire, & Hennon (2002) used point-light displays to test for sensitivity to path and manner with an intermodal preferential looking paradigm. Three-year-olds were able to match a motion, stripped of any identifying information other than path and manner, with the target verb spoken by an experimenter. A follow-up experiment indicated that young children could also produce appropriate (action) verbs when prompted, using only point-light displays. The authors concluded that point-light displays are (and will be in future research) useful for detecting the components most useful to verb learning. Yet before being able to learn a verb that encodes manner or path, it is conceivable that the infant should attend to such components in an event. Zheng & Goldin-Meadow (2002) provided preliminary evidence that manner and path are attended to even with little to no previous exposure to language models. For more recent accounts, see Casasola, Bhagwat, & Ferguson (2006), Choi (2006a; 2006b), and Pulverman, Hirsh-Pasek, & Golinkoff (2006).

The manipulation of parts of a motion event, involving an interaction between two objects such as using a Michottian manipulation with varied velocities and/or delays, in relation to verb usage and word choice has not been studied to-date. While the original elements described by Talmy as constituting a motion event, such as Path and Manner, should be addressed, they may only be determinants of verb meaning for 'simple' motion events (i.e. events involving only one agent, not involving an interaction with some recipient). More components may be involved in whole-body interactions that should not be overlooked. In P. R. Cohen's Maps for Verbs (1998) framework, elements such as velocity and energy transfer serve as candidates for other elements accessible in the semantic core.

## 11.4  Maps for Verbs

We tested the hypothesis that word choices are conditioned on patterns of motion in a study called 'Maps for Verbs'. We began with a dynamical representation of verbs that denote physical interactions between two agents or objects named A and B. Examples include 'bump', 'hit', 'push', 'overtake', 'chase', 'follow', 'harass', 'hammer', 'shove', 'meet', 'touch', 'propel', 'kick', and 'bounce' (P. R. Cohen 1998).

The Maps for Verbs framework proposes that simple interactions between whole bodies can be characterized by the physical dynamics of the interaction.

According to the framework, whole-body interactions are naturally divided into three phases: before, during, and after contact. Figure 11.1 depicts these three phases. A given interaction is then described as a trajectory through these phases. Maps enable identification of characteristic patterns present in the dynamics of classes of interactions.

P. R. Cohen (1998) proposes that the Before and After phases should plot relative velocity against the distance between the two bodies. Relative velocity is the difference between the velocity of one body, A, and another, B. Many verbs (e.g. transitive verbs) predicate one body as the 'actor' and the other as the 'target' (or 'subject' or 'recipient') of the action. For example, in a scenario involving a PUSH, the actor is the one doing the pushing, and the target is the body being pushed. By convention, the actor is designated as body A and the target is body B. Thus, when relative velocity is positive, the actor's velocity is greater than that of the target; and when relative velocity is negative, the target's velocity is greater than that of the actor. Distance, in turn, is the measure of the distance between the bodies.

The During phase plots perceived energy transfer (from the actor to the target) against time or distance. If energy transfer is positive, then the actor is imparting to the target more energy than the target originally had; if energy transfer is negative, then the situation is reversed: the target is imparting more energy to the actor. To measure perceived energy transfer, we used the simplification of calculating the acceleration of the actor in the direction of the target while in contact.

Figure 11.1 depicts a set of labeled trajectories that characterize the component phases of seven interaction types as described by the verbs 'push', 'shove', 'hit', 'harass', 'bounce', 'counter-shove', and 'chase'. Using these labels, an interaction can be described as a triple of trajectory labels, indicating the Before, During, and After characteristic trajectories. For example, [**b,b,b**] describes a 'shove': The actor approaches the target at a greater velocity than the target,

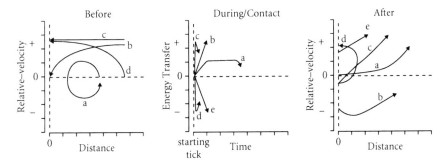

FIGURE 11.1. Maps for Verbs model of the three phases of interaction

closing the distance between the two bodies. As it nears the target, the actor slows, decreasing its velocity to match that of the target. Trajectory **b** of the Before phase in Figure 11.1 illustrates these dynamics. At contact, the relative velocity is near or equal to zero. During the contact phase, the actor rapidly imparts more energy to the target in a short amount of time, as illustrated by **b** of the During phase. And after breaking off contact with the target, the agent rapidly decreases its velocity while the target moves at a greater velocity from the energy imparted it (trajectory **b** in the After phase).

Following this scheme, the remaining six interaction types are characterized by the following triples:

'push': **b, a, a**: Begins like 'shove', but at contact relative velocity is near or equal to zero and the actor smoothly imparts more energy to the target; after breaking contact, the agent gradually decreases its velocity.

'hit': **c** or **d, c, c**: May begin with the actor already at high velocity relative to the target or increasing in relative velocity, and thus is characterized by **c** or **d** in the Before phase.

'harass': **c** or **d, c, d**: Similar to a hit, except the After phase involves the actor quickly recovering its speed and moving back toward the target, not allowing the distance between the two to get very large. 'Harass' highlights that all interactions are not to be viewed only as single movement to contact, but may involve many such movements to contact, one after another, and may even switch between different kinds of contact interaction.

'bounce': **c** or **d, d, e**: Along with 'counter-shove', 'bounce' involves the target making a more reactive response to the actor's actions. 'Bounce' begins like a 'hit' or 'harass', but at contact, the target transfers a large amount of energy back to the actor.

'counter-shove': **b** or **c** or **d, e, e**: A version of a 'shove' where the target imparts energy to the actor.

'chase': **a, -, -**: The agent moves toward the target, closing the distance between the two, but never quite making contact, so the during and after phases are not relevant. This is depicted as the circular trajectory **a** in the Before phase.

Morrison, Cannon, & Cohen (2004) used these seven classes of interaction as the basis for a study in which we looked at the frequency of verb usage of adults asked to describe the interaction types after observing them. Forty-four undergraduates (M = 20.5 years old) at the University of Massachusetts participated in this study. We used *breve* 1.4, an environment for developing realistic multi-body simulations in a three dimensional world with physics (Klein 2002), to implement a model of the seven interaction classes described in the previous section. The model is rendered as two generic objects (a blue ball for the actor and a red ball for the target) moving on a white background.

We generated a set of movies based on the rendered interactions. For several of the interaction classes we also varied the behavior of the target object, as follows: the target object, (a) did not move except when contacted ('stationary'), (b) moved independently in a random walk ('wander'), or (c) moved according to billiard-ball ballistic physics, based on the force of the collision ('coast'). We generated a total of 17 unique movies. These were presented on a G3 iMac with 14-inch screen.

A total of 18 movies were presented to each participant, with 'chase' being viewed twice. After watching a movie, participants were asked to write down answers to questions on a sheet of paper given to them by the experimenter. The questions were the same for every movie:

1. What are the balls doing in this movie? (Give your overall impression of what was happening between them, the 'gist'.)
2. What is the red ball doing?
3. What is the blue ball doing?
4. Can you think of any words to describe the tone or the mood of the movie? (e.g. the balls are friendly/not friendly.)

The experimenter encouraged participants to write as much as they could to describe the movies. All the action words and other content words for each trial were extracted and 'canonicalized', converting verbs in different tenses or forms (ending in '-ed', '-ing', etc.) to a unique form. Also, negation phrases, such as 'it's not zooming' or 'red didn't move', were also transformed into a single token, e.g. not-zooming and not-moving.

After canonicalization, we kept only the verbs from the content words (a total of 155 verbs). The following 65 verbs are those that were each used by ten or more subjects to describe the movies:

advancing, annoying, approaching, attaching, attacking, avoiding, backing, beating, bouncing, bullying, bumping, catching, charging, chasing, circling, coming, controlling, defending, dominating, escaping, fighting, floating, following, forcing, getting, giving, guiding, helping, hitting, kissing, knocking, leading, leaving, letting, looking, losing, nudging, pursuing, placing, playing, propelling, pushing, repeating, repelling, resisting, responding, rolling, running, shoving, slamming, slowing, sneaking, standing, standing one's ground, staying, stopping, striking, tagging, teasing, touching, traveling, trying, waiting, wanting, winning

Recall that the Maps for Verbs framework hypothesizes that a representation based on the dynamics of Before, During, and After interactions are a foundation for the semantics of verbs describing physical interactions between objects. If this hypothesis is correct, we would expect the subjects in the preceding experiment to use particular verbs when describing the movies they observed. Furthermore, movies that share the same kind of dynamics in terms of Before, During, and After phases of interaction should elicit similar groups of verbs. To see whether this was the case, we clustered the 17 movies according the frequency of word usage, where frequency was according to the number of different subjects who used a given word to describe a movie (i.e. if five different subjects used the word 'approaching' to describe the 'harass'-'wander movie', then the frequency recorded was 5). We used hierarchical agglomerative clustering (Duda, Hart, & Stork 2001) to cluster the movies based on these word frequencies. Figure 11.2 shows the generated dendrogram tree depicting the results of clustering (ignore for the moment the additional labels and notation to the right).

At first the dendrogram looks disappointing; while there is some structure, it is not clear how to interpret the groupings. However, recall that the movies were generated by behavioral programs, written in *breve*, that attempt to match the dynamics outlined in Figure 11.1. The program specifications do not guarantee that the salient perceptual features of Before, During, and After interaction dynamics will be perspicuous.

To explore this further, we independently observed each movie and chose what we believed to be features that help distinguish movies from one another. We came up with a total of five very simple features:

'purpose before', 'purpose after': whether red (the target of the interaction) looked purposeful before or after contact ('purposeful' was in terms of whether red appeared to change its heading on its own);

'reactive during': whether red seemed to react to contact ('react' was in terms of whether red appeared to change its behavior based on blue's contact);

'gentle start', 'gentle end': whether the initial or final stages of the contact appeared gentle.

We then went through each movie and assigned a minus or plus, depending on whether each feature was present ('−' = no; '+' = yes). Some cases were uncertain, so we assigned a '+?' or '−?'; and some cases were indeterminable, receiving a '?'. We have placed these feature vectors next to the corresponding leaves of the dendrogram in Figure 11.2. We can now see that there is significant structure to the clusters, based on the similar features that are grouped. The internal node labeled 1 in the dendrogram tree of Figure 11.2 distinguishes the cluster of movies where red is not reactive to blue's contact while the contact begins gently from movies in which red is reactive and contact does not begin

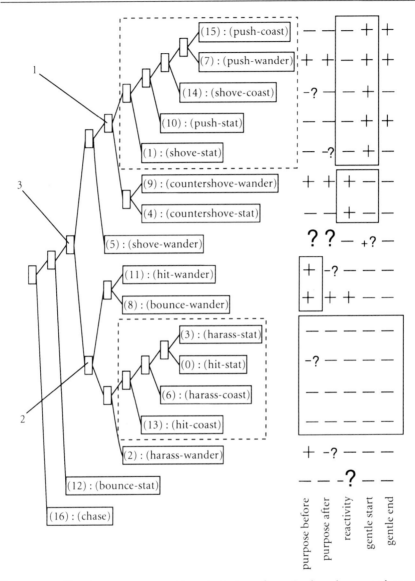

FIGURE 11.2. Dendrogram representing clustering of movies based on word usage frequencies, where word usage is based on the number of different subjects who used a given word. The complete set of 155 verbs were used to characterize word usage. The labels inside the leaves of the dendrogram correspond to movie names; the numbers are unique identifiers assigned by the clustering procedure and should be ignored.

gently. The node labeled 2 in the dendrogram distinguishes between whether red looks purposeful before or after interaction (although the placement of 'harass'-'wander' is problematic; it should be associated with 'hit'-'wander' and 'bounce'-'wander'). Finally, the node labeled 3 appears to separate groups of movies that involve gentle starts to interactions or red reactivity from movies that all involve abrupt starts and ends to the contact phase of interaction (except for 'bounce'-'wander').

These results indicate that the dynamical features present in the movies influence the choice of verbs used by the subjects to describe the movies. Although, to date, we have only tested a subset of the possible interaction types outlined in Figure 11.1, the data thus far seem to indicate that the distinctions in the Maps for Verbs framework, which led us to develop 17 distinct movies, do in fact influence word choices people make. We have demonstrated that words are selected preferentially in response to different dynamics, but we have not demonstrated that the distinctions in the Maps for Verbs framework (i.e. the different paths through the three phases) are systematically associated with different distributions of evoked words. Word use certainly seems to be associated with dynamics, but not necessarily exclusively to the ways described by the maps for verbs framework. More work is needed to show that this framework *predicts* distributions of word use for different movies. Preliminary work with preschool-aged children suggests that even for fairly new English language learners, word choice is associated with these motion components of an interaction. While there may be other elements and parameters contributing to **s**, this study suggests that we have a good starting place to begin looking seriously at the detail of dynamics involved, their development, and also the possibility of additional elements involved in giving these whole-body interactions meaning.

### 11.4.1 *Future directions*

There are two avenues of research within the existing Maps for Verbs framework which could make considerable advancements to our understanding of motion-based semantics and word choice. Given the evidence discussed throughout in this chapter, both cross-cultural and developmental work in this area is warranted.

If we were to test a Korean population with manipulations set out in the Maps for Verbs framework, would we see the same distributions of verbs for the movies? At this point we can only speculate. Not only have differences been found in verb usage between Korean and English speakers (e.g. Choi & Bowerman, 1991) but also differences in spatial categorization (Choi, McDonough, Bowerman, & Mandler 1999), and (potentially) universal early sensitivities to these distinctions may disappear if the language does not lexicalize them (e.g. McDonough, Choi, & Mandler 2003). It would be interesting to see along which parameters the Korean population categorizes whole-body interactions in comparison to an English-speaking population. Furthermore, we know nothing, at this point, about cross-cultural emphases on different phases of an interaction. It is plausible that dynamics within some phases of an interaction dictate word choice more than others. And maybe these phase differences vary across cultures. In other words, perhaps the sensitivity of one language is focused on the Before phase—the behavior of an agent, just prior to contact with another, has more influence over the semantics, and therefore word choice, than whatever happens in the During or After Contact phases. In another language, events within the During Contact phase might be most informative. Comparing the phases would not only be informative in discovering something more about cross-cultural ontological distinctions (and similarities), but might also suggest other contributing elements to **s**, present in the semantic core.

As we have also discussed in this chapter, infants are remarkably capable of extracting meaning from motion dynamics. The work described earlier in this chapter on the perception of physical causality in infancy suggests that infants may make categorical distinctions along the dimensions of the Maps for Verbs framework within the first year of life. Perhaps, as they learn the interaction categories most relevant to the language being learned, we will see a loss of some distinctions and the refinement of others. Perhaps infants' early sensitivities to motion dynamics also contribute new elements to the semantic core from which **s** is formed.

## 11.5 Concluding remarks

We began this chapter with the question of how we can effectively communicate through language in an ever-changing world. We suggested that, in a world that is constantly in motion, movement must be a powerful cue for extracting meaningful information from our environment. A general theory of word meaning was offered, stating that, in all language, words uttered are conditioned on the representation of a situation, which is largely made up of these situational motion elements we perceive. We reviewed the literature that even unobservable elements in **s** can be inferred through motion. Moreover, we provided a review of cognitive and linguistic evidence suggesting that infants are initially sensitive to far more motion cues than are later represented in **s**, and that whether or not the sensitivity remains will depend on how the perceptual system and spoken language bundles these elements. The context provides meaning, and while we claim motion as a central component, we have never claimed it as the sole contributor. We have reviewed several proposed representational semantic frameworks for investigating motion elements and discussed one in detail, Maps for Verbs. However, there may be other additional motion elements that have not yet been discovered.

We interact with objects in the world from birth, so it seems fitting to study the dynamics of interactions when making claims about semantics and language development. But other potential contributors to s should also be examined, such as syntax, other words uttered, intention, and number. While these domains are studied extensively on their own, a comprehensive associative learning theory would have to consider the influences of all of the elements that may contribute to s in order to build a complete model for the acquisition of word meaning.

# References

Acredolo, L. (1979) Laboratory versus home: the effect of the environment on the 9-month-old infant's choice of spatial reference system. *Developmental Psychology* 15: 666–7.

——Adams, A., & Goodwyn, S. W. (1984) The role of self-produced movement and visual tracking in infant spatial orientation. *Journal of Experimental Child Psychology* 38: 312–27.

Adolph, K. E., Eppler, M. A., & Gibson, E. J. (1993) Crawling versus walking: infants' perception of affordances for locomotion on slopes. *Child Development* 64: 1158–74.

Aleksander, I. (1973) Random logic nets: stability and adaptation. *International Journal of Man Machine Studies* 5: 115–31.

Alibali, M. W. (1999) How children change their minds: strategy change can be gradual or abrupt. *Developmental Psychology* 35: 127–45.

Allport, A. (1990) Visual attention. In M. Posner (ed.), *Foundations of Cognitive Science*. Cambridge, MA: MIT Press, 631–82.

Altmann, G., & Kamide, Y. (2004) Now you see it, now you don't: mediating the mapping between language and the visual world. In J. Henderson and F. Ferreira (eds.), *The Interaction of Vision, Language, and Action*. New York: Psychology Press, 347–86.

Alvarez, G. A., & Franconeri, S. (2007) How many objects can you track? Evidence for a resource-limited attentive tracking mechanism. *Journal of Vision* 7: 14.1–10.

Amari, S. (1977) Dynamics of pattern formation in lateral-inhibition type neural fields. *Biological Cybernetics* 27: 77–87.

——(1989) Dynamical stability of formation of cortical maps. In M. A. Arbib & S. Amari (eds.), *Dynamic Interactions in Neural Networks: Models and Data*. New York: Springer 15–34.

——& Arbib, M. A. (1977) Competition and cooperation in neural nets. In J. Metzler (ed.), *Systems Neuroscience*. New York: Academic Press 119–65.

Anderson, J. A., Silverstein, J. W., Ritz, S. A., & Jones, R. S. (1977) Distinctive features, categorical perception, and probability learning: some applications of a neural model. *Psychological Review* 84: 413–51.

Anderson, J. R. (2000) *Cognitive Psychology and its Implications*. New York: Worth.

Ansari, D., & Karmiloff-Smith, A. (2002) Atypical trajectories of number development: a neuroconstructivist perspective. *Trends in Cognitive Sciences* 6: 511–16.

Antrobus, J. S., Antrobus, J. S., & Singer, J. L. (1964) Eye movements accompanying daydreaming, visual imagery, and thought suppression. *Journal of Abnormal and Social Psychology* 69: 244–52.

Arterberry, M. E., & Bornstein, M. H. (2001) Three-month-old infants' categorization of animals and vehicles based on static and dynamic attributes. *Journal of Experimental Child Psychology* 80: 333–46.

——— (2002) Infant perceptual and conceptual categorization: the roles of static and dynamic attributes. *Cognition* 86: 1–24.

Ashley, A., & Carlson, L. A. (2007) Interpreting proximity terms involves computing distance and direction. *Language and Cognitive Processes* 22: 1021–44.

Avraamides, M. (2003) Spatial updating of environments described in texts. *Cognitive Psychology* 47: 402–31.

Awh, E., Jonides, J., & Reuter-Lorenz, P. A. (1998) Rehearsal in spatial working memory. *Journal of Experimental Psychology: Human Perception and Performance* 24: 780–90.

——— Smith, E. E., Buxton, R. B., Frank, L. R., Love, T., & Wong, E. (1999) Rehearsal in spatial working memory: evidence from neuroimaging. *Psychological Science* 10: 433–7.

Baddeley, A. D. (1986) *Working Memory*. Oxford: Oxford University Press.

Bailey, D. (1997) When push comes to shove: a computational model of the role of motor control in the acquisition of action verbs. Doctoral dissertation, Computer Science Division, University of California, Berkeley.

Baillargeon, R. (1993) The object concept revisited: new direction in the investigation of infants' physical knowledge. In C. Granrud (ed.), *Visual Perception and Cognition in Infancy*. Hillsdale, NJ: Erlbaum, 265–315.

Baldwin, D. A. (1991) Infants' contribution to the achievement of joint reference. *Child Development* 62: 875–90.

—— (1993) Early referential understanding: infants' ability to recognize referential acts for what they are. *Developmental Psychology* 29: 832–43.

—— & Baird, J. A. (2001) Discerning intentions in dynamic human action. *Trends in Cognitive Sciences* 5: 171–8.

——— Saylor, M. M., & Clark, M. A. (2001) Infants parse dynamic actions. *Child Development* 72: 708–18.

—— Markman, E. M., Bill, B., Desjardins, R. N., Irwin, J. M., & Tidball, G. (1996) Infants' reliance on a social criterion for establishing word–object relations. *Child Development* 67: 3135–53.

Ball, D. (1992) Magical hopes: manipulatives and the reform of math education. *American Educator* 16: 14–18.

Ballard, D. H., Hayhoe, M., & Pelz, J. (1995) Memory representations in natural tasks. *Journal of Cognitive Neuroscience* 7: 68–82.

————Pook, P. K., & Rao, R. P. N. (1997) Deictic codes for the embodiment of cognition. *Behavioral and Brain Sciences* 20: 723–67.

———— & Yu, C. (2003) A multimodal learning interface for word acquisition. *Proceedings of the International Conference on Acoustics, Speech and Signal Processing* (Hong Kong) 5: 784–7.

Barlow, H. (1972) Single units and sensation: a neuron doctrine for perceptual psychology. *Perception* 1: 371–94.

Baron-Cohen, S. (1994) How to build a baby that can read minds: cognitive mechanisms in mind-reading. *Cahiers de Psychologie Cognitive* 13: 513–52.

Baroody, A. J. (1989) One point of view: manipulatives don't come with guarantees. *Arithmetic Teacher* 37: 4–5.

————(1990) How and when should place value concepts and skills be taught? *Journal for Research in Mathematics Education* 21: 281–6.

Barsalou, L. W. (1999) Perceptual symbol systems. *Behavioral and Brain Sciences* 22: 577–660.

————(2005) Abstraction as a dynamic construal in perceptual symbol systems. In L. Gershkoff-Stowe & D. H. Rakinson (eds.), *Building Object Categories in Developmental Time*. Hillsdale, NJ: Erlbaum, 389–431.

Bartsch, K., & Wellman, H. M. (1995) *Children Talk about the Mind*. Oxford: Oxford University Press.

Bastian, A., Riehle, A., Erlhagen, W., & Schöner, G. (1998) Prior information preshapes the population representation of movement direction in motor cortex. *NeuroReport* 9: 315–19.

———— Schöner, G., & Riehle, A. (2003) Preshaping and continuous evolution of motor cortical representations during movement preparation. *European Journal of Neuroscience* 18: 2047–58.

Beach, K. (1988) The role of external mnemonic symbols in acquiring an occupation. In M. M. Gruneberg & R. N. Sykes (eds.), *Practical Aspects of Memory*, vol. 1. New York: Wiley, 342–6.

Behrmann, M. (2000) Spatial reference frames and hemispatial neglect. In M. Gazzaniga (ed.), *The New Cognitive Neurosciences*, 2nd edn. Cambridge, MA: MIT Press, 651–66.

Bellugi, U., Marks, S., Bihrle, A., & Sabo, H. (1988) Dissociation between language and cognitive functions in Williams syndrome. In B. D. Bishop & K. Mogford (eds.), *Language Development in Exceptional Circumstances*. Hillsdale, NJ: Lawrence Erlbaum, 177–89.

Bergen, B., Lindsay, S., Matlock, T., & Narayanan, S. (2007) Spatial and linguistic aspects of visual imagery in sentence comprehension. *Cognitive Science* 31: 733–64.

———— Narayan, S., & Feldman, J. (2003) Embodied verbal semantics: evidence from an image-verb matching task. In *Proceedings of the 25th Annual Conference of the Cognitive Science Society*. Mahwah, NJ: Erlbaum, 139–44.

Berry, D. S., & Springer, K. (1993) Structure, motion, and preschoolers perception of social causality. *Ecological Psychology* 5: 273–83.

Bertenthal, B. I. (1993) Infants' perception of biomechanical motions: intrinsic image and knowledge-based constraints. In C. E. Granrud (ed.), *Visual Perception and Cognition in Infancy* Hillsdale, NJ: Erlbaum, 175–214.

——Campos, J., & Barrett, K. (1984b) Self-produced locomotion: an organizer of emotional, cognitive, and social development in infancy. In R. Emde and R. Harmon (eds.), *Continuities and Discontinuities*. New York: Plenum Press 175–210.

——& Pinto, J. (1994) Global processing of biological motions. *Psychological Science* 5: 221–5.

——Proffitt, D. R., & Cutting, J. E. (1984a) Infant sensitivity to figural coherence in biomechanical motions. *Journal of Experimental Child Psychology* 37: 213–30.

Bloom, P. (2000) *How Children Learn the Meanings of Words*. Cambridge, MA: MIT Press.

Bollt, E. M., Stanford, T., Lai, Y., & Zyczkowski, K. (2000) Validity of threshold-crossing analysis of symbolic dynamics from chaotic time series. *Physical Review Letters* 85: 3524–7.

Bornstein, M. H., & Korda, N. O. (1984) Discrimination and matching within and between hues measured by reaction times: some implications for categorical perception and levels of information processing. *Psychological Research* 46(3): 207–22.

——— (1985) Identification and adaptation of hue: parallels in the operation of mechanisms that underlie categorical perception in vision and in audition. *Psychological Research* 47(1): 1–17.

Boroditsky, L. (2000) Metaphoric structuring: understanding time through spatial metaphors. *Cognition* 75: 1–28.

—— (2001) Does language shape thought? English and Mandarin speakers' conceptions of time. *Cognitive Psychology* 43: 1–22.

——& Ramscar, M. (2002) The roles of body and mind in abstract thought. *Psychological Science* 13: 185–8.

Bower, G. H., & Morrow, D. G. (1990) Mental models in narrative comprehension. *Science* 247: 44–8.

Brandt, S. A., & Stark, L. W. (1997) Spontaneous eye movements during visual imagery reflect the content of the visual scene. *Journal of Cognitive Neuroscience* 9: 27–38.

Bremner, J. G. (1978) Egocentric versus allocentric spatial coding in nine-month-old infants: factors influencing the choice of a code. *Developmental Psychology* 14: 346–55.

Brent, M. R., & Cartwright, T. A. (1996) Distributional regularity and phonotactic constraints are useful for segmentation. *Cognition* 61: 93–125.

Bridgeman, B., Gemmer, A., Forsman, T., & Huemer, V. (2000) Processing spatial information in the sensorimotor branch of the visual system. *Vision Research* 40: 3539–52.

——Peery, S., & Anand, S. (1997) Interaction of cognitive and sensorimotor maps of visual space. *Perception and Psychophysics* 59: 456–69.

Brown, J., Johnson, M., Paterson, S., Gilmore, R., Longhi, E., & Karmiloff-Smith, A. (2003) Spatial representation and attention in toddlers with Williams syndrome and Down syndrome. *Neuoropsychologia* 41: 1037–46.

Brown, P. F., Pietra, S., Pietra, V., & Mercer, R. L. (1993) The mathematics of statistical machine translation: parameter estimation. *Computational Linguistics* 19: 263–311.

Brown, P., & Levinson, S. C. (1993) Uphill and downhill in Tzeltal. *Journal of Linguistic Anthropology* 3: 46–74.

Bruner, J. S. (1961) The act of discovery. *Harvard Educational Review*, 31, pp. 21–32.

—— (1965) The growth of mind. *American Psychologist* 20: 1007–17.

—— (1966) *Toward a Theory of Instruction.* New York: Norton.

—— (1990) *Acts of Meaning.* Cambridge, MA: Harvard University Press.

Bub, D. N., Masson, M. E. J., & Bukach, C. M. (2003) Gesturing and naming: the use of functional knowledge in object identification. *Psychological Science* 14: 467–72.

Butterworth, G. (1991) The ontogeny and phylogeny of joint visual attention. In A. Whiten (ed.), *Natural Theories of Mind: Evolution, Development, and Simulation of Everyday Mindreading.* Oxford: Blackwell, 223–32.

Carey, S. (2001) Whorf versus continuity theorists: bringing data to bear on the debate. In M. Bowerman & S. C. Levinson (eds.), *Language Acquisition and Conceptual Development.* New York: Cambridge University Press 185–214.

Carlson, L. A. (forthcoming) Parsing space around objects. In V. Evans and P. Chilton (eds.), *Language, Cognition and Space: The State of the Art and New Directions.* London: Equinox.

—— & Covey, E. S. (2005) How far is near? Inferring distance from spatial descriptions. *Language and Cognitive Processes* 20: 617–32.

—— & Kenny, R. (2006) Interpreting spatial terms involves simulating interactions. *Psychonomic Bulletin and Review* 13: 682–8.

—— & Logan, G. D. (2001) Using spatial terms to select an object. *Memory and Cognition* 29: 883–92.

—— —— (2005) Attention and spatial language. In L. Itti, G. Rees, & J. Tsotsos (eds.), *Neurobiology of Attention.* New York: Academic Press, 330–36.

—— & van Deman, S. R. (2004) The space in spatial language. *Journal of Memory and Language* 51: 418–36.

Carlson-Radvansky, L. A., Covey, E. S., & Lattanzi, K. M. (1999) 'What' effects on 'where': functional influences on spatial relations. *Psychological Science* 10: 516–21.

—— & Irwin, D. E. (1993) Frames of reference in vision and language: where is above? *Cognition* 46: 223–44.

—— —— (1994) Reference frame activation during spatial term assignment. *Journal of Memory and Language* 33: 646–71.

—— & Jiang, Y. (1998) Inhibition accompanies reference frame selection. *Psychological Science* 9: 386–91.

—— & Logan, G. D. (1997) The influence of reference frame selection on spatial template construction. *Journal of Memory and Language* 37: 411–37.

—— & Tang, Z. (2000) Functional influences on orienting a reference frame. *Memory and Cognition* 28: 812–20.

Casad, E. H. (1988) Conventionalization of Cora locationals. In B. Rudzka-Ostyn (ed.), *Topics in Cognitive Linguistics*, vol. 50. Amsterdam: Benjamins, 345–78.

Casasola, M., Bhagwat, J., & Ferguson, K. T. (2006) Precursors to verb learning: infants' understanding of motion events. In K. Hirsh-Pasek & R. M. Golinkoff (eds.), *Action Meets Word: How Children Learn Verbs*. Oxford: Oxford University Press, 160–90.

Casey, M. P. (1996) The dynamics of discrete time computation with application to recurrent neural networks and finite state machine extraction. *Neural Computation* 8: 1135–78.

Cheng, K. (1986) A purely geometric module in the rat's spatial representation. *Cognition* 23: 149–78.

—— & Newcombe, N. S. (2005) Is there a geometric module for spatial orientation? Squaring theory and evidence. *Psychonomic Bulletin and Review* 12: 1–23.

Chi, M. T. H., Feltovich, P. J., & Glaser, R. (1981) Categorization and representation of physics problems by experts and novices. *Cognitive Science* 5: 121–52.

—— & Koeske, R. (1983) Network representation of a child's dinosaur knowledge. *Developmental Psychology* 19: 29–39.

Choi, S. (2006a) Influence of language-specific input on spatial cognition: categories of containment. *First Language* 26: 207–32.

—— (2006b) Preverbal spatial cognition and language-specific input: categories of containment and support. In K. Hirsh-Pasek & R. M. Golinkoff (eds.), *Action Meets Word: How Children Learn Verbs*. Oxford: Oxford University Press, 191–207.

—— & Bowerman, M. (1991) Learning to express motion events in English and Korean: the influence of language-specific lexicalization patterns. *Cognition* 41: 83–121.

—— McDonough, L., Bowerman, M., & Mandler, J. M. (1999) Early sensitivity to language-specific spatial categories in English and Korean. *Cognitive Development* 14: 241–68.

Chun, M. M., & Wolfe, J. M. (2001) Visual attention. In E. B. Goldstein (ed.), *Blackwell Handbook of Perception*. Malden, MA: Blackwell, 272–310.

Churchland, P., & Sejnowski, T. (1992) *The Computational Brain*. Cambridge, MA: MIT Press.

Clark, A. (1997) *Being There: Putting Brain, Body and World Together Again*. Cambridge, MA: MIT Press.

—— (1998) Magic words: how language augments human computation. In P. Carruthers & J. Boucher (eds.), *Language and Thought: Interdisciplinary Themes*. Cambridge: Cambridge University Press, 162–83.

—— (1999) An embodied cognitive science? *Trends in Cognitive Sciences* 3: 345–51.

—— (2003) *Natural-Born Cyborgs: Minds, Technologies, and the Future of Human Intelligence*. Oxford: Oxford University Press.

Clark, H. (1973) Space, time, semantics, and the child. In T. E. Moore (ed.), *Cognitive Development and the Acquisition of Language*. New York: Academic Press, 27–64.

—— (1992) *Arenas of Language Use*. Chicago: University of Chicago Press.

—— (1996) *Using Language*. Cambridge: Cambridge University Press.

—— & Chase, W. G. (1972) On the process of comparing sentences against pictures. *Cognitive Psychology* 3: 472–517.

Clausner, T. C., & Croft, W. (1999) Domains and image schemas. *Cognitive Linguistics* 10: 1–31.

Clearfield, M., Dineva, E., Smith, L. B., Fiedrich, F., & Thelen, E. (2007) Cue salience and infant preseverative reaching. MS.

Cleeremans, A., Servan-Schreiber, D., & McClelland, J. L. (1989) Finite state automata and simple recurrent networks. *Neural Computation* 1: 372–81.

Cohen, J. D., Braver, T. S., & O'Reilly, R. C. (1996) A computational approach to prefrontal cortex, cognitive control, and schizophrenia: recent developments and current challenges. *Philosophical Transactions of the Royal Society London, B: Biology* 351: 1515–27.

Cohen, L. B., & Amsel, G. (1998) Precursors to infants' perception of the causality of a simple event. *Infant Behavior and Development* 21: 713–32.

—— & Oakes, L. M. (1993) How infants perceive a simple causal event. *Developmental Psychology* 29: 421–33.

—— Rundell, L. J., Spellman, B. A., & Cashon, C. H. (1999) Infants' perception of causal chains. *Psychological Science* 10: 412–18.

Cohen, P. R. (1998) Maps for verbs. In *Proceedings of the Information and Technology Systems Conference*, 15th IFIP World Computer Congress, Vienna, 21–33.

—— Oates, T., Adams, N., & Beal, C. R. (2001) Robot baby 2001. *Lecture Notes in Artificial Intelligence* 2225: 32–56. Washington, DC: IEEE Press.

—— —— Beal, C. R., & Adams, N. (2002) Contentful mental states for RobotBaby. Paper presented at meeting of the American Association for Artificial Intelligence, Edmonton.

Coltheart, M. (1981) The MRC psycholinguistic database. *Quarterly Journal of Experimental Psychology* 33A: 497–505.

Compte, A., Brunel, N., Goldman-Rakic, P. S., & Wang, X. J. (2000) Synaptic mechanisms and network dynamics underlying spatial working memory in a cortical network model. *Cerebral Cortex* 10: 910–23.

Cooper, R. (1974) The control of eye fixation by the meaning of spoken language. *Cognitive Psychology* 6: 84–107.

Coulson, S., & Matlock, T. (2001) Metaphor and the space structuring model. *Metaphor and Symbol* 16: 295–316.

Coventry, K. R., & Garrod, S. C. (2004) *Saying, Seeing and Acting: The Psychological Semantics of Spatial Prepositions*. Hove, UK: Psychology Press.

—— Prat-Sala, M., & Richards, L. (2001) The interplay between geometry and function in the comprehension of over, under, above and below. *Journal of Memory and Language* 44: 376–98.

Craver-Lemley, C., & Arterberry, M. E. (2001) Imagery induced interference on a visual detection task. *Spatial Vision* 14: 101–19.

Crawford, L. E., Regier, T., & Huttenlocher, J. (2000) Linguistic and non-linguistic spatial categorization. *Cognition* 75: 209–35.

Creem, S. H., & Profitt, D. R. (2001) Grasping objects by their handles: a necessary interaction between cognition and action. *Journal of Experimental Psychology: Human Perception and Performance* 27: 218–28.

Crutchfield, J. P. (1994) The calculi of emergence: computation, dynamics and induction. *Physica D* 75: 11–54.

Csibra, G., Gergely, G., Biro, S., Koos, O., & Brockbank, M. (1999) Goal-attribution without agency cues: the perception of 'pure reason' in infancy. *Cognition* 72: 237–67.

Dale, R., & Spivey, M. (2005) From apples and oranges to symbolic dynamics: a framework for conciliating notions of cognitive representation. *Journal of Experimental and Theoretical Artificial Intelligence* 17: 317–42.

Damasio, A. R. (1989) The brain binds entitites and events by multiregional activation from convergence zones. *Neural Computation* 1: 123–32.

—— & Damasio, H. (1994) Cortical systems for retrieval of concrete knowledge: the convergence zone framework. In C. Koch & J. L. Davis (eds.), *Large-Scale Neuronal Theories of the Brain: Computational Neuroscience*. Cambridge, MA: MIT Press, 61–74.

Damper, R. I., & Harnad, S. R. (2000) Neural network models of categorical perception. *Perception and Psychophysics* 62: 843–67.

Deloache, J. S. (2000) Dual representation and children's use of scale models. *Child Development* 71: 329–38.

Demarais, A. M., & Cohen, B. H. (1998) Evidence for image scanning eye movements during transitive inference. *Biological Psychology* 49: 229–47.

Dempster, A. P., Laird, N. M., & Rubin, D. B. (1977) Maximum likelihood from incomplete data via the em algorithm. *Journal of the Royal Statistical Society* 39: 1–38.

Denis, M., & Kosslyn, S. M. (1999) Scanning visual mental images: a window on the mind. *Cahiers de Psychologie Cognitive/Current Psychology of Cognition* 18: 409–65.

Dennett, D. (1996) *Kinds of Minds*. New York: Basic Books.

Devaney, R. L. (2003) *An Introduction to Chaotic Dynamical Systems* 2nd edn. Boulder, CO: Westview Press.

Diamond, A. (1990a) The development and neural bases of memory functions as indexed by the AB and delayed response tasks in human infants and infant monkeys. In A. Diamond (ed.), *The Development and Neural Bases of Higher Cognitive Functions*. New York: NewYork Academy of Sciences, 637–76.

—— (1990b) Developmental time course in human infants and infant monkeys and the neural bases of inhibitory control in reaching. In A. Diamond (ed.), *The Development and Neural Bases of Higher Cognitive Functions*. New York: New York Academy of Sciences, 637–76.

—— (1998) Understanding the A-not-B error: working memory vs. reinforced response, or active vs. latent trace. *Developmental Science* 1: 185–9.

Diedrich, F. J., Highlands, T. M., Spahr, K. A., Thelen, E., & Smith, L. B. (2001) The role of target distinctiveness in infant perseverative reaching. *Journal of Experimental Child Psychology* 78: 263–90.

Dienes, Z. P. (1960) *Building Up Mathematics.* London: Hutchinson.

Dietrich, E., & Markman, A. B. (2003) Discrete thoughts: why cognition must use discrete representations. *Mind and Language* 18: 95–119.

Dorman, M. F. (1974) Auditory evoked potential correlates of speech sound discrimination. *Perception and Psychophysics* 15: 215–20.

Douglas, R., & Martin, K. (1998) Neocortex. In G. M. Shepherd (ed.), *The Synaptic Organization of the Brain.* New York: Oxford University Press, 459–509.

Dourish, P. (2001) *Where The Action Is: The Foundations of Embodied Interaction.* Cambridge, MA: MIT Press.

Duda, R. O., Hart, P. E., & Stork, D. G. (2001) *Pattern Classification.* New York: Wiley.

Edelman, S. (1999) *Representation and Recognition in Vision.* Cambridge, MA: MIT Press.

Eilan, N., McCarthy, R., & Brewer, B. (1993) *Spatial Representation.* Oxford: Blackwell.

Ellis, R., & Tucker, M. (2000) Micro-affordance: the potentiation of components of action by seen objects. *British Journal of Psychology* 91: 451–71.

Elman, J. L. (1991) Distributed representations, simple recurrent networks, and grammatical structure. *Machine Learning* 7: 195–224

—— Bates, E. A., Johnson, M. H., Karmiloff Smith, A., Parisi, D., & Plunkett, K. (1996) *Rethinking Innateness: A Connectionist Perspective on Development.* Cambridge, MA: MIT Press.

Emmorey, K., & Casey, S. (1995) A comparison of spatial language in English and American Sign Language. *Sign Language Studies* 88: 255–88.

Engebretson, P. H., & Huttenlocher, J. (1996) Bias in spatial location due to categorization: comment on Tversky and Schiano. *Journal of Experimental Psychology: General* 125: 96–108.

Engel, A. K., Koenig, P., Kreiter, A. K., Schillen, T. B., & Singer, W. (1992) Temporal coding in the visual cortex: new vistas on integration in the nervous system. *Trends in Neurosciences* 15: 218–26.

Erickson, M. A., & Kruschke, J. K. (2002) Rule based extrapolation in perceptual categorization. *Psychonomic Bulletin and Review* 9: 160–68.

Erlhagen, W., Bastian, A., Jancke, D., Riehle, A., & Schöner, G. (1999) The distribution of neuronal population activation (DPA) as a tool to study interaction and integration in cortical representations. *Journal of Neuroscience Methods* 94: 53–66.

Ervin-Tripp, S. (1973) Some strategies for the first two years. In T. Moore (ed.), *Cognitive Development and the Acquisition of Language.* New York: Academic Press, 261–86.

Farah, M. J. (1985) Psychophysical evidence for a shared representational medium for mental images and percepts. *Journal of Experimental Psychology* 114: 91–103.

—— (1989) Mechanisms of imagery perception interaction. *Journal of Experimental Psychology: Human Perception and Performance* 15: 203–11.

Farah, M. J., Brunn, J. L., Wong, A. B., Wallace, M. A., & Carpenter, P. A. (1990) Frames of reference for allocating attention to space: evidence from the neglect syndrome. *Neuropsychologia* 28: 335–47.

Fennema, E. (1972) The relative effectiveness of a symbolic and a concrete model in learning a selected mathematical principle. *Journal for Research in Mathematics Education* 3: 233–8.

Fillmore, C. J. (1971) *Santa Cruz Lectures on Deixis*. Bloomington, IA: University of Indiana Linguistics Club.

Finke, R. A. (1985) Theories relating mental imagery to perception. *Psychological Bulletin* 98: 236–59.

Flavell, J. H. (1999) Cognitive development: children's knowledge about the mind. *Annual Review of Psychology* 50: 21–45.

Fodor, J. A. (1983) *The Modularity of Mind*. Cambridge, MA: MIT Press.

Franklin, N., & Henkel, L. A. (1995) Parsing surrounding space into regions. *Memory and Cognition* 23: 397–407.

—————— & Zangas, T. (1995) Parsing surrounding space into regions. *Memory and Cognition* 23: 397–407.

Freeman, W. J. (2000) *Neurodynamics:An Exploration in Mesoscopic Brain Dynamics*. London: Springer.

Friedman, M. (1978) The manipulative materials strategy: the latest pied piper? *Journal for Research in Mathematics Education* 9: 78–80.

Fuson, K. C. (1988) *Children's Counting and Concepts of Number*. New York: Springer.

—————— & Briars, D. J. (1990) Using a base-ten blocks learning/teaching approach for first- and second-grade place-value and multidigit addition and subtraction. *Journal for Research in Mathematics Education* 21: 180–206.

Gapp, K-P. (1994) *A Computational Model of the Basic Meanings of Graded Composition Spatial Relations in 3-D Space* (Tech. Rep. III). Saarbrücken: Dept of Computer Science, Universität des Saarlandes.

—————— (1995) Angle, distance, shape, and their relationship to projective relations. In J. D. Moore & J. F. Lehman (eds.), *Proceedings of the 17th Annual Conference of the Cognitive Science Society*. Mahwah, NJ: Cognitive Science Society, 112–17.

Garnham, A. (1989) A unified theory of the meaning of some spatial relational terms. *Cognition* 31: 45–60.

Gauthier, I., & Logothetis, N. K. (2000) Is face recognition not so unique after all? *Cognitive Neuropsychology* 17: 125–42.

Gelman, R. (1991) Epigenetic foundations of knowledge structures: initial and tran-scendent constructions. In S. Carey & R. Gelman (eds.), *Epigenesis of Mind: Essays on Biology and Cognition*. Hillsdale, NJ: Erlbaum 293–322.

—————— & Baillargeon, R. (1983) A review of some Piagetian concepts. In J. H. Flavell and E. Markman (eds.), *Handbook of Child Development*, vol. 3: *Cognitive Development* New York: Wiley, 167–230.

Gentner, D. (1978) On relational meaning: the acquisition of verb meaning. *Child Development* 49: 988–98.

—— (1982) Why nouns are learned before verbs: linguistic relativity versus natural partitioning. In S. A. Kuczaj, II (ed.), *Language Development*, vol. 2: *Language, Thought, and Culture*. Hillside, NJ: Erlbaum, 301–34.

—— & Rattermann, M. J. (1991) Language and the career of similarity. In S. A. Gelman & J. P. Byrnes (eds.), *Perspectives on Language and Thought: Interrelations in Development*. Cambridge: Cambridge University Press, 225–77.

Georgopoulos, A. (1995) Motor cortex and cognitive processing. In M. Gazzaniga (ed.), *The Cognitive Neurosciences*. Cambridge, MA: MIT Press, 507–12.

—— Georgopoulos, A. P., Kurz N., & Landau, B. (2004) Figure copying in Williams syndrome and normal subjects. *Experimental Brain Research* 157: 137–46.

—— Kalaska, J. F., Caminiti, R., & Massey, J. T. (1982) On the relations between the direction of two dimensional arm movements and cell discharge in primate motor cortex. *Journal of Neuroscience* 2: 1527–37.

—— Schwartz, A. B., & Kettner, R. E. (1986) Neuronal population coding of movement direction. *Science* 223: 1416–19.

Gergely, G., Nadasdy, Z., Csibra, G. & Biro, S. (1995) Taking the intentional stance at 12 months of age. *Cognition* 56: 165–93.

Gibbs, R. W. (1994) *The Poetics of Mind: Figurative Thought, Language, and Understanding*. New York: Cambridge University Press.

—— (1996) Why many concepts are metaphorical. *Cognition* 61: 309–19.

—— (2006) *Embodiment and Cognitive Science*. New York: Cambridge University Press.

—— & Colston, H. L. (1995) The cognitive psychological reality of image schemas and their transformations. *Cognitive Linguistics* 6, 347–78.

—— Ström, L. K., & Spivey-Knowlton, M. J. (1997) Conceptual metaphors in mental imagery for proverbs. *Journal of Mental Imagery* 21: 83–109.

Gillette, J., Gleitman, H., Gleitman, L., & Lederer, A. (1999) Human simulations of vocabulary learning. *Cognition* 73: 135–76.

Ginsburg, H. P. (1977) The psychology of arithmetic thinking. *Journal of Children's Mathematical Behavior* 1: 1–89.

—— & Golbeck, S. L. (2004) Thoughts on the future of research on mathematics and science learning and education. *Early Childhood Research Quarterly* 19: 190–200.

Gleitman, L. (1990) The structural sources of verb meanings. *Language Acquisition* 1: 1–55.

—— & Newport, E. (1995) The invention of language by children: environmental and biological influences on the acquisition of language. In L. Gleitman & M. Liberman (eds.), *An Invitation to Cognitive Science: Language*, vol. 1. Cambridge, MA: MIT Press, 1–24.

Glenberg, A. M. (1997) What is memory for? *Behavioral and Brain Sciences* 20: 1–55.

—— & Kaschak, M. P. (2002) Grounding language in action. *Psychonomic Bulletin and Review* 9: 558–65.

—— Robertson, D. A., Kaschak, M. P., & Malter, A. J. (2003) Embodied meaning and negative priming. *Behavioral and Brain Sciences* 26: 644–8.

Goldstone, R. L., & Sakamoto, Y. (2003) The transfer of abstract principles governing complex adaptive systems. *Cognitive Psychology* 46: 414–66.

——& Son, J. Y. (2005) The transfer of scientific principles using concrete and idealized simulations. *Journal of the Learning Sciences* 14: 69–110.

——Chung, H. L., Hirsh-Pasek, K., Liu, J., Bertenthal, B. L., Brand, R., Maguire, M. J., & Hennon, E. (2002) Young children can extend motion verbs to point light displays. *Developmental Psychology* 38: 604–14.

Golinkoff, R. M., & Kerr, J. L. (1978) Infants' perception of semantically defined action role changes in filmed events. *Merrill Palmer Quarterly* 24: 53–61.

Gopnik, A., & Choi, S. (1995) Names, relational words, and cognitive development in English and Korean speakers: nouns are not always learned before verbs. In M. Tomasello & W. E. Merriman (eds.), *Beyond Names for Things: Young Children's Acquisition of Verbs*. Hillsdale, NJ: Erlbaum, 63–80.

Gouteux, S., Thinus-Blanc, C., & Vauclair, J. (2001) Rhesus monkeys use geometric and non-geometric information during a reorientation task. *Journal of Experimental Psychology: General* 130: 505–19.

——Vauclair, J., & Thinus-Blanc, C. (2001) Reorientation in a small-scale environment by 3-, 4-, and 5-year-old children. *Congnitive Development* 16: 853–69.

Greeno, J. G. (1989) Situations, mental models and generative knowledge. In D. Klahr & K. Kotovsky (eds.), *Complex Information Processing: The Impact of Herbert Simon*. Hillsdale, NJ: Erlbaum 285–318.

Gregory, R. (1981) *Mind in Science: A History of Explanations in Psychology and Physics*. London: Weidenfeld & Nicolson.

Griffin, Z. M. (2004) Why look? reasons for eye movements related to language production. In J. Henderson & F. Ferreira (eds.), *The Integration of Language, Vision, and Action: Eye Movements and the Visual World*. New York: Taylor & Francis, 213–47.

——Bock, K. (2000) What the eyes say about speaking. *Psychological Science* 11: 274–9.

Gruber, O., & Goschke, T. (2004) Executive control emerging from dynamic interactions between brain systems mediating language, working memory, and attentional processes. *Acta Psychologica* 115: 105–21.

Gupta, P., & MacWhinney, B. (1997) Vocabulary acquisition and verbal short-term memory: computational and neural bases. *Brain and Language* 59: 267–333.

————Feldman, H. M., & Sacco, K. (2003) Phonological memory and vocabulary learning in children with focal lesions. *Brain and Language* 87: 241–52.

Guylai, E. (2000) Attribution of meaning from movement of simple objects. *Perceptual and Motor Skills* 90: 27–35.

Hainstock, E. (1978) *The Essential Montessori*. New York: Plume.

Harnad, S. (ed.) (1987) *Categorical Perception: The Groundwork of Cognition*. New York: Cambridge University Press.

——(1990) The symbol grounding problem. *Physica D* 42: 335–46.

Hartley, T., Burgess, N., Lever, C., Cacucci, F., & O'Keefe, J. (2000) Modeling place fields in terms of the cortical inputs to the hippocampus. *Hippocampus* 10: 369–79.

Hatano, G. (1982) Cognitive consequences of practice in culture specific procedural skills. *Quarterly Newsletter of the Laboratory of Comparative Human Cognition* 4: 15–17.

Hauser, M. D., Chomsky, N., & Fitch, W. T. (2002) The faculty of language: what is it, who has it, and how did it evolve? *Science* 298: 1569–79.

Hayward, W. G., & Tarr, M. J. (1995) Spatial language and spatial representation. *Cognition* 55: 39–84.

Heidegger, M. (1961[1927]) *Being and Time*, trans. J. Macquarie & E. Robinson. Oxford: Blackwell. [First published New York: Harper & Row.]

Heider, F., & Simmel, M. (1944) An experimental study of apparent behavior. *American Journal of Psychology* 57: 243–59.

Hermer, L., & Spelke, E. (1994) A geometric process for spatial reorientation in young children. *Nature* 370: 57–9.

——— (1996) Modularity and development: the case of spatial reorientation. *Cognition* 61: 195–232.

Herskovits, A. (1986) *Language and Spatial Cognition: An Interdisciplinary Study of the Prepositions of English*. Cambridge: Cambridge University Press.

Hiebert, J., & Carpenter, T. P. (1992) Learning and teaching with understanding. In D. A. Grouws (ed.), *Handbook of Research on Mathematics Teaching and Learning*. New York: Macmillan, 65–97.

Hoff, E. (2003) The specificity of environmental influence: socioeconomic status affects early vocabulary development via maternal speech. *Child Development* 74: 1368–78.

Hoffman, J. E., & Landau, B. (in preparation) Visual short-term memory in people with Williams syndrome.

——— & Pagani, B. (2003) Spatial breakdown in spatial construction: evidence from eye fixations in children with Williams syndrome. *Cognitive Psychology* 46: 260–301.

Hummel, J. E. (2001) Complementary solutions to the binding problem in vision: implications for shape perception and object recognition. *Visual Cognition* 8: 489–517.

—— & Holyoak, K. J. (2003) A symbolic connectionist theory of relational inference and generalization. *Psychological Review* 110: 220–64.

Humphreys, G. W., & Riddoch, M. J. (2003) From 'what' to 'where': neuropsychological evidence for implicit interactions between object- and space-based attention. *Psychological Science* 14: 487–92.

Hund, A. M., & Plumert, J. M. (2002) Delay-induced bias in children's memory for location. *Child Development* 73: 829–40.

——— (2003) Does information about what things are influence children's memory for where things are? *Developmental Psychology* 39: 939–48.

—— Plumert, J. M., & Benney, C. J. (2002) Experiencing nearby locations together in time: the role of spatiotemporal contiguity in children's memory for location. *Journal of Experimental Child Psychology* 82: 200–225.

Hupach, A., & Nadel, L. (2005) Reorientation in a rhombic environment: no evidence for an encapsulated geometric model. *Cognitive Development* 20: 279–302.

—— & Spencer, J. P. (2003) Developmental changes in the relative weighting of geometric and experience-dependent location cues. *Journal of Cognition and Development* 4: 3–38.

Huttenlocher, J., Hedges, L. V., & Duncan, S. (1991) Categories and particulars: prototype effects in estimating spatial location. *Psychological Review* 98: 352–76.

—— —— & Vevea, J. L. (2000) Why do categories affect stimulus judgment? *Journal of Experimental Psychology: General* 129: 220–41.

Huttenlocher, J., Newcombe, N. S., & Sandberg, E. H. (1994) The coding of spatial location in young children. *Cognitive Psychology* 27: 115–48.

—— & Presson, C. C. (1973) Mental rotation and the perspective problem. *Cognitive Psychology* 4: 277–99.

—— —— (1979) The coding and transformation of spatial information. *Cognitive Psychology* 11: 375–94.

—— & Vasilyeva, M. (2003) How toddlers represent enclosed spaces. *Cognitive Science* 27: 749–66.

Imai, M., Gentner, D., & Uchida, N. (1994). Children's theories of word meaning: The role of shape similarity in early acquisition. *Cognitive Development*, 9, 45–75.

Ifrah, G. (1981) *The Universal History of Numbers: From Prehistory to the Invention of the Computer*, 2nd edn 2000. New York: Wiley.

Jackendoff, R. (1983) *Semantics and Cognition*. Cambridge, MA: MIT Press.

—— (1992) *Languages of the Mind*. Cambridge, MA: MIT Press.

—— (1996) The architecture of the linguistic-spatial interface. In P. Bloom, M. A. Peterson, L. Nadel, & M. F. Garrett (eds.), *Language and Space*. Cambridge, MA: MIT Press, 1–30.

—— (2002) *Foundations of Language: Brain, Meaning, Grammar, Evolution*. Oxford: Oxford University Press.

Jarrold, C., Baddeley, A. D., & Hewes, A. K. (1999) Genetically dissociated components of working memory: evidence from Down's and Williams syndrome. *Neuropsychologia* 37: 637–51.

Johansson, G. (1973) Visual perception of biological motion and a model for its analysis. *Perception and Psychophysics* 14: 201–11.

Johnson, J. S., Spencer, J. P., & Schöner, G. (in press) Moving to higher ground: the dynamic field theory and the dynamics of visual cognition. In F. Garzón, A. Laakso, & T. Gomila (eds.), *Dynamics and Psychology*, special issue of *New Ideas in Psychology*.

Johnson, M. (1987) *The Body in the Mind*. Chicago: University of Chicago Press.

Johnson-Laird, P. N. (1983) *Mental Models: Towards a Cognitive Science of Language, Inference, and Consciousness*. Cambridge: Cambridge University Press.

—— (1998) Imagery, visualization, and thinking. In J. Hochberg (ed.), *Perception and Cognition at Century's End: Handbook of Perception and Cognition* 2nd edn. San Diego, CA: Academic Press, 441–67.

Jones, G. A., & Thornton, C. A. (1993) Children's understanding of place value: a framework for curriculum development and assessment. *Young Children* 48: 12–18.

Jordan, H., Reiss, J. E., Hoffman, J. E., & Landau, B. (2002) Intact perception of biological motion in the face of profound spatial deficits: Williams syndrome. *Psychological Science* 13: 162–7.

Kahneman, D., Treisman, A. and Gibbs, B. (1992) The reviewing of object files: object-specific integration of information. *Cognitive Psychology* 24: 175–219.

Kamii, C. (1986) Place value: an explanation of its difficulty and educational implications for the primary grades. *Journal of Research in Childhood Education* 1: 75–86.

—— & Joseph, L. (1988) Teaching place value and double-column addition. *Arithmetic Teacher* 35: 48–52.

Kaminski, J. A., Sloutsky, V. M., & Heckler, A. F. (2005) Relevant concreteness and its effects on learning and transfer. In B. Bara, L. Barsalou, & M. Bucciarelli (eds.), *Proceedings of the 27th Annual Conference of the Cognitive Science Society*.

———— (2006a) Effects of concreteness on representation: an explanation for differential transfer. In R. Sun & N. Miyake (eds.), *Proceedings of the 28th Annual Conference of the Cognitive Science Society*.

———— (2006b) Do children need concrete instantiations to learn an abstract concept? In R. Sun & N. Miyake (eds.), *Proceedings of the 28th Annual Conference of the Cognitive Science Society*.

Kan, I. P., Barsalou, L. W., Solomon, K. O., Minor, J. K., & Thompson-Schill, S. L. (2003) Role of mental imagery in a property verification task: fMRI evidence for perceptual representations of conceptual knowledge. *Cognitive Neuropsychology* 20: 525–40.

Kaufman, A. S., & Kaufman, N. L. (1990) *Kaufman Brief Intelligence Test*. Circle Pines, MN: American Guidance Service.

Kelso, J. A. S. (1994) *Dynamic Patterns: The Self Organization of Brain and Behavior*. Cambridge, MA: MIT Press.

Kim, J. (1998) *Mind in a Physical World*. Cambridge, MA: MIT Press.

Kirlik, A. (1998) Everyday life environments. In W. Bechtel & G. Graham (eds.), *A Companion to Cognitive Science*. Oxford: Blackwell, 702–12.

Kirsh, D. (1995) The intelligent use of space. *Artificial Intelligence* 73(1–2): 31–68.

—— and Maglio, P. (1994) On distinguishing epistemic from pragmatic action. *Cognitive Science* 18: 513–49.

Klein, J. (2002) Breve: a 3D simulation environment for the simulation of decentralized systems and artificial life. In *Proceedings of Artificial Life VIII, the 8th International Conference on the Simulation and Synthesis of Living Systems*. http://www.spiderland.org/breve/.

Kluender, K. R., Diehl, R. L., & Killeen, P. R. (1987) Japanese quail can learn phonetic categories. *Science* 237: 1195–7.

Kohonen, T. (1982) Self-organized formation of topologically correct feature maps. *Biological Cybernetics* 43: 59–69.

Kosslyn, S. M., Ball, T. M., & Reiser, B. J. (1978) Visual images preserve metric spatial information: evidence from studies of image scanning. *Journal of Experimental Psychology: Human Perception and Performance* 4: 47–60.

—— Thompson, W. L., Kim, I. J., & Alpert, N. M. (1995) Topographical representations of mental images in primary visual cortex. *Nature* 378: 496–8.

Kosslyn, S. M., Wraga, M., & Alpert, N. M. (2001) Imagining rotation by endogenous versus exogenous forces: distinct neural mechanisms. *Neuroreport* 12: 2519–25.

Kotovsky, L., & Gentner, D. (1996) Comparison and categorization in the development of relational similarity. Child Development 67: 2797–2822.

Kuelpe, O. (1902) Über die objectivirung und subjectivirung von sinneseindrucken [On objective and subjective sensory impressions]. *Philosophische Studien* 19: 508–56.

Kuhl, P. K., & Miller, J. D. (1975) Speech perception by the chinchilla: voiced voiceless distinction in alveolar plosive consonants. *Science* 190: 69–72.

——Tsao, F.-M., & Liu, H.-M. (2003) Foreign-language experience in infancy: effects of short-term exposure and social interaction on phonetic learning. *PNAS* 100: 9096–9101.

Laeng, B., & Teodorescu, D. S. (2002) Eye scanpaths during visual imagery reenact those of perception of the same visual scene. *Cognitive Science* 26: 207–31.

Lakoff, G. (1987a) *Women, Fire and Dangerous Things: What Categories Reveal about the Mind*. Chicago, IL: University of Chicago Press.

——& Johnson, M. (1980a) *Metaphors We Live By*. Chicago, IL: University of Chicago Press.

————(1980b) The metaphorical structure of the human conceptual system. *Cognitive Science* 4: 195–208.

————(1999) *Philosophy in the Flesh: The Embodied Mind and its Challenge to Western Thought*. New York: Basic Books.

——& Nunez, R. E. (2000) *Where Mathematics Comes From: How the Embodied Mind Brings Mathematics into Being*. New York: Basic Books.

Lakusta, L., & Landau, B. (2005) Starting at the end: the importance of goals in spatial language. *Cognition* 96: 1–33.

Land, M., & Mcleod, P. (2000) From eye movements to actions: how batsmen hit the ball. *Nature Neuroscience* 3: 1340–45.

——Mennie, N., & Rusted, J. (1999) The roles of vision and eye movements in the control of activities of daily living. *Perception* 28: 1311–28.

Landau, B. (2000) Language and space. In B. Landau, J. Sabini, & J. Jonides (eds.), *Perception, Cognition, and Language: Essays in Honor of Henry and Lila Gleitman*. Cambridge, MA: MIT Press, 209–90.

——Hoffman, J. E., & Kurz, N. (2006) Object recognition with severe spatial deficits in Williams syndrome: sparing and breakdown. *Cognition* 100: 483–510.

——& Jackendoff, R. (1993) 'What' and 'where' in spatial language and spatial cognition. *Behavioral and Brain Sciences* 16: 217–65.

——& Kurz, N. (2004) Spatial patterns and their transformations. Unpublished data.

——& Zukowski, A. (2003) Objects, motions, and paths: spatial language in children with Williams syndrome. In C. Mervis (ed.), special issue on Williams syndrome, *Developmental Neuropsychology* 23: 105–37.

Landau, K. B., Smith, L. B., & Jones, S. S. (1988) The importance of shape in early lexical learning. *Cognitive Development* 3: 299–321.

Langacker, R. W. (1987) *Foundations of Cognitive Grammar*, vol. 1: *Theoretical Pre-requisites*. Stanford, CA: Stanford University Press.

——(1990) *Foundations of Cognitive Grammar*, vol. 2: *Descriptive Application*. Stanford, CA: Stanford University Press.

——(1991) *Concept, Image, Symbol: The Cognitive Basis of Grammar*. Berlin: Mouton de Gruyter.

——(1993) Grammatical traces of some 'invisible' semantic constructs. *Language Sciences* 15: 323–55.

——(2000) Virtual reality. *Studies in the Linguistic Sciences* 29: 77–103.

——(2002) A study in unified diversity: English and Mixtec locatives. In N. J. Enfield (ed.), *Ethnosyntax: Explorations in Grammar and Culture*. Oxford: Oxford University Press, 138–61.

Lashley, K. S. (1951) The problem of serial order in behavior. In L. A. Jeffress (ed.), *Cerebral Mechanisms in Behavior*. New York: Wiley.

Learmonth, A. E., Nadel, L., & Newcombe, N. S. (2002) Children's use of landmarks: implications for modularity theory. *Psychological Science* 13: 337–41.

——Newcombe, N., & Huttenlocher, J. (2001) Toddlers' use of metric information and landmarks to reorient. *Journal of Experimental Child Psychology* 80: 225–44.

Legerstee, M., Barna, J., & DiAdamo, C. (2000) Precursors to the development of intention at 6 months: understanding people and their actions. *Developmental Psychology* 36: 627–34.

Leslie, A. M. (1982) Perception of causality in infants. *Perception* 11: 173–86.

——(1984) Spatiotemporal continuity and the perception of causality in infants. *Perception* 13: 287–305.

——(1988) The necessity of illusion: perception and thought in infancy. In L. Weiskrantz (ed.), *Thought Without Language*. Oxford: Clarendon Press, 185–210.

——(1994) ToMM, ToBY, and agency: core architecture and domain specificity. In L. A. Hirschfeld & S. A. Gelman (eds.), *Mapping the Mind: Domain Specificity in Cognition and Culture*. Cambridge: Cambridge University Press, 119–48.

——& Keeble, S. (1987) Do six-month-olds perceive causality? *Cognition* 25: 265–88.

——Xu, F., Tremoulet, P., & Scholl, B. (1998) Indexing and the object concept: developing 'what' and 'where' systems. *Trends in Cognitive Science* 2: 10–18.

Lesperance, Y., & Levesque, H. J. (1990) Indexical knowledge in robot plans. In *Proceedings of the 8th National Conference on Artificial Intelligence (AAAI-90)*, Menlo Park, CA: AAAI Press/MIT Press, 1030–37.

Lettvin, J. Y. (1995) J. Y. Lettvin on grandmother cells. In M. Gazzaniga (ed.), *The Cognitive Neurosciences*. Cambridge, MA: MIT Press, 434–5.

Levelt, W. J. M. (1984) Some perceptual limitations on talking about space. In A. J. van Doorn, W. A. van der Grind, & J. J. Koenderink (eds.), *Limits in Perception*. Utrecht: VNU Science Press, 323–58.

——(1996) Perspective taking and ellipsis in spatial descriptions. In P. Bloom, M. A. Peterson, L. Nadel, & M. Garrett (eds.), *Language and Space*. Cambridge, MA: MIT Press, 77–108.

Levinson, S. (1994) Vision, shape, and linguistic description: Tzeltal body-part terminology and object description. Special issue on 'Spatial conceptualization in Mayan languages', *Linguistics* 32: 791–855.

——(1996) Frames of reference and Molyneux's questions: cross-linguistic evidence. In P. Bloom, M. A. Peterson, L. Nadel, & M. Garrett (eds.), *Language and Space.* Cambridge, MA: MIT Press, 109–69.

——(2003) *Space in Language and Cognition.* Cambridge: Cambridge University Press.

Liberman, A. M. (1982) On finding that speech is special. *American Psychologist* 37: 148–67.

——Harris, K. S., Hoffman, H. S., & Griffith, B. C. (1957) The discrimination of speech sounds within and across phoneme boundaries. *Journal of Experimental Psychology* 54: 358–68.

————Kinney, J. A., & Lane, H. (1961) The discrimination of relative onset time of the components of certain speech and nonspeech patterns. *Journal of Experimental Psychology* 61: 379–88.

Lillard, A. S. (2005) *Montessori: The Science Behind the Genius.* New York: Oxford University Press.

Lipinski, J., Sandamirskaya, Y., & Schöner, G. (2009c) Swing it to the left, swing it to the right: Enacting flexible spatial language using a neurodynamic framework. MS in preparation.

——Spencer, J. P., & Samuelson, L. K. (2009a) Linguistic and non-linguistic spatial cognitive systems travel together through time. MS in preparation.

————(2009b) Active spatial terms bias active spatial memory: metric dynamic interactions between linguistic and non-linguistic spatial processes. MS in preparation.

Logan, G. D. (1994) Spatial attention and the apprehension of spatial relations. *Journal of Experimental Psychology: Human Perception and Performance* 20: 1015–36.

——(1995) Linguistic and conceptual control of visual spatial attention. *Cognitive Psychology* 28: 103–74.

——& Compton, B. J. (1996) Distance and distraction effects in the apprehension of spatial relations. *Journal of Experimental Psychology: Human Perception and Performance* 22: 159–72.

——& Sadler, D. D. (1996) A computational analysis of the apprehension of spatial relations. In P. Bloom, M. A. Peterson, L. Nadel, & M. F. Garrett (eds.), *Language and Space.* Cambridge, MA: MIT Press, 493–529.

Loomis, J. M., Lippa, Y., Klatzky, R. L., & Golledge, R. G. (2002) Spatial updating of location specified by 3-D sound and spatial language. *Journal of Experimental Psychology: Learning, Memory, and Cognition* 28: 335–45.

Lourenco, S. F., & Huttenlocher, J. (2006) How do young children determine location? Evidence from disorientation tasks. *Cognition* 100: 511–29.

————(2007) Using geometry to specify location: implications for spatial coding in children and nonhuman animals. *Psychological Research* 71: 252–64.

——————— & Vasilyeva, M. (2005) Toddlers' representations of space: the role of viewer perspective. *Psychological Science* 16: 255–9.

Love, B. C., Medin, D. L., & Gureckis, T. M. (2004) SUSTAIN: a network model of category learning. *Psychological Review* 111: 309–32.

Luck, S. J., & Vogel, E. K. (1997) The capacity of visual working memory for features and conjunctions. *Nature* 390(6657): 279–81.

Lund, K., & Burgess, C. (1996) Producing high dimensional semantic spaces from lexical co-occurrence. *Behavior Research Methods, Instruments and Computers* 28: 203–8.

Lupyan, G., & Spivey, M. J. (in press) Perceptual processing is facilitated by ascribing meaning to novel stimuli. *Current Biology.*

Mack, N. K. (1993) Learning rational numbers with understanding: the case of informal knowledge. In T. P. Carpenter, E. Fennema, & T. A. Romberg (eds.), *Rational Numbers: An Integration of Research.* Hillsdale, NJ: Erlbaum, 85–105.

—— (2000) Long-term effects of building on informal knowledge in a complex content domain: the case of multiplication of fractions. *Journal of Mathematical Behavior* 19: 307–32.

—— (2001) Building on informal knowledge through instruction in a complex content domain: partitioning, units, and understanding multiplication of fractions. *Journal for Research in Mathematics Education* 32: 267–95.

Macnamara, J. (1982) *Names for Things: A Study of Child Language.* Cambridge, MA: MIT Press.

MacWhinney, B. (1989) Competition and lexical categorization. In R. Corrigan, F. Eckman, & M. Noonan (eds.), *Linguistic Categorization.* Amsterdam: Benjamins, 195–242.

—— & Snow, C. (1985) The child language data exchange system. *Journal of Child Language* 12: 271–96.

Mainwaring, S., Tversky, B., Ohgishi, M., & Schiano, D. (2003) Descriptions of simple spatial scenes in English and Japanese. *Spatial Cognition and Computation* 3: 3–42.

Mandler, J. (1992) How to build a baby II: conceptual primitives. *Psychological Review* 99: 587–604.

—— (2000) Perceptual and conceptual processes in infancy. *Journal of Cognition and Development* 1: 3–36.

—— & McDonough, L. (1996) Drinking and driving don't mix: inductive generalization in infancy. *Cognition* 59: 307–35.

Marcus, G. F. (2001) *The Algebraic Mind: Integrating Connectionism and Cognitive Science.* Cambridge, MA: MIT Press.

—— Vouloumanos, A., & Sag, I. A. (2003) Does Broca's play by the rules? *Nature Neuroscience* 6: 652–3.

Markman, E. M. (1989). *Categorization and naming in children: Problems of induction.* Cambridge, MA: The MIT Press.

Markson, L., & Bloom, P. (1997) Evidence against a dedicated system for word learning in children. *Nature* 385(6619): 813–15.

Massaro, D. W. (1987) *Speech Perception by Ear and Eye: A Paradigm for Psychological Inquiry.* Hillsdale, NJ: Erlbaum.

—— (1998) *Perceiving Talking Faces: From Speech Perception to a Behavioral Principle.* Cambridge, MA: MIT Press.

Matlock, T. (2004) Fictive motion as cognitive simulation. *Memory and Cognition* 32, 1389–1400.

—— (in press) Fictive motion as cognitive simulation. *Memory and Cognition.*

—— Ramscar, M., & Boroditsky, L. (2005) The experiential link between spatial and temporal language. *Cognitive Science* 29: 655–64.

McClelland, J. L., McNaughton, B. L., & O'Reilly, R. C. (1995) Why there are complementary learning systems in the hippocampus and neocortex: insights from the successes and failures of connectionist models of learning and memory. *Psychological Review* 102: 419–57.

McCulloch, W. (1965) *Embodiments of Mind.* Cambridge, MA: MIT Press.

McDonough, L., Choi, S., & Mandler, J. (2003) Understanding spatial relations: flexible infants, lexical adults. *Cognitive Psychology* 46: 229–59.

McGlone, M. S., & Harding, J. L. (1998) Back (or forward?) to the future: the role of perspective in temporal language comprehension. *Journal of Experimental Psychology: Learning Memory and Cognition* 24: 1211–23.

McGurk, H., & MacDonald, J. W. (1976) Hearing lips and seeing voices. *Nature* 264: 746–8.

McMurray, B., Tanenhaus, M. K., Aslin, R. N., & Spivey, M. J. (2003) Probabilistic constraint satisfaction at the lexical/phonetic interface: evidence for gradient effects of within category VOT on lexical access. *Journal of Psycholinguistic Research* 32: 77–97.

McMurrer, J. (2007) Choices, Changes and Challenges: Curriculum and Instruction in the NCLB Era. Report from the Center of Education Policy (24 July). Washington, D.C.

McTaggart, J. (1908) The unreality of time. *Mind* 17: 457–74.

Menninger, K. (1969) *Number Words and Number Symbols: A Cultural History of Numbers.* New York: Dover.

Merleau-Ponty, M. (1968) *Phenomenology of Perception.* London: Routledge & Kegan Paul.

Mervis, C. B., Morris, C. A., Bertrand, J., & Robinson, B. (1999) Williams syndrome: findings from an integrated program of research. In H. Tager-Flusberg (ed.), *Neurodevelopmental Disorders: Contributions to a New Framework from the Cognitive Neurosciences.* Cambridge, MA: MIT Press, 65–110.

Meyer, A. S., Sleiderink, A. M., & Levelt, W. J. (1998) Viewing and naming objects: eye movements during noun phrase production. *Cognition* 66: B25–B33.

Michotte, A. (1963) *The Perception of Causality.* London: Methuen.

Miller, G. A., & Johnson-Laird, P. N. (1976) *Language and Perception.* Cambridge, MA: Harvard University Press.

Miller, S. P. & Mercer, C. D. (1993). Using data to learn concrete-semiconcreteabstract instruction for students with math disabilities. *Learning Disabilties Research & Practice*, 8, 89–96.

Miura, I. T., Okamotot, Y., Kim, C. C., Steere, M., & Fayol, M. (1993) First graders' cognitive representation of number and understanding of place value: cross-national comparisons—France, Japan, Korea, Sweden and the United States. *Journal of Educational Psychology* 85: 24–30.

Mix, K. S. (1999) Similarity and numerical equivalence: appearances count. *Cognitive Development* 14: 269–97.

—— Huttenlocher, J., & Levine, S. C. (1996) Do preschool children recognize auditory-visual numerical correspondences? *Child Development* 67: 1592–1608.

—— —— —— (2002a) Multiple cues for quantification in infancy: is number one of them? *Psychological Bulletin* 128: 278–94.

—— —— —— (2002b). *Quantitative Development in Infancy and Early Childhood.* New York: Oxford University Press.

Molfese, D. (1987) Electrophysiological indices of categorical perception for speech. In S. Harnad (ed.), *Categorical Perception: The Groundwork of Cognition.* New York: Cambridge University Press.

Montessori, M. (1917) *The Advanced Montessori Method.* New York: Stokes.

—— (1964) *The Montessori Method.* New York: Schocken Books.

Moore, C. M., Yantis, S., & Vaughan, B. (1998) Object-based visual selection: evidence from perceptual completion. *Psychological Science* 9: 104–10.

Morrison, C. T., Cannon, E. N., & Cohen, P. R. (2004) When push comes to shove: a study of the relation between interaction dynamics and verb use. In *Working Notes of AAAI Spring Symposium Workshop: Language Learning: An Interdisciplinary Perspective.* Menlo Park, CA: AAAI.

Morrow, D. G., & Clark, H. H. (1988) Interpreting words in spatial descriptions. *Language and Cognitive Processes* 3: 275–91.

Moyer, P. S. (2001) Are we having fun yet? How teachers use manipulatives to teach mathematics. *Educational Studies in Mathematics* 47: 175–97.

Munakata, Y. (1998) Infant perseveration and implications for object permancence theories: a PDP model of the A-not-B task. *Developmental Science* 1: 161–84.

—— & McClelland, J. L. (2003) Connectionist models of development. *Developmental Science* 6: 413–29.

Munhall, K. G., & Vatikiotis-Bateson, E. (1998) The moving face during speech communication. In R. Campbell & B. Dodd (eds.), *Hearing by Eye II: Advances in the Psychology of Speechreading and Auditory Visual Speech.* Hove, UK: Psychology Press, 123–39.

Murphy, G. (1996) On metaphoric representation. *Cognition* 60: 173–204.

Naigles, L. R., Eisenberg, A. R., Kako, E. T., Highter, M., & McGraw, N. (1998) Speaking of motion: verb use in English and Spanish. *Language and Cognitive Processes* 13: 521–49.

Nelson, T. O., & Chaiklin, S. (1980) Immediate memory for spatial location. *Journal of Experimental Psychology: Human Learning and Memory* 6: 529–45.

Newcombe, N., Huttenlocher, J., Drummey, A. B., & Wiley, J. G. (1998) The development of spatial location coding: place learning and dead reckoning in the second and third years. *Cognitive Development* 13: 185–200.

Newcombe, N. S., & Huttenlocher, J. (2000) *Making Space: The Development of Spatial Representation and Reasoning.* Cambridge, MA: MIT Press.

Noice, H., & Noice, T. (2001) Learning dialogue with and without movement. *Memory and Cognition* 29: 820–27.

Norman, D. (1993) *Things That Make Us Smart.* Cambridge, MA: Perseus Books.

O'Hearn, K., & Landau, B. (2007) Mathematical skill in individuals with Williams syndrome: evidence from a standardized mathematics battery. *Brain and Cognition* 64: 238–46.

————— & Hoffman, J. E. (2005a) Subitizing in people with Williams syndrome and normally developing children. Poster presented at the biennial meeting of the Society for Research in Child Development, Atlanta, GA.

————— (2005b) Multiple object tracking in people with Williams syndrome and in normally developing children. *Psychological Science* 16: 905–12.

O'Hearn Donny, K., landau, B., Courtney, S., & Hoffman, J. E. (2004) Working memory for location and identity in Williams syndrome. Poster presented at the 4th annual conference for the Vision Sciences Society, Sarasota, FL (April–May).

O'Keefe, J. (2003) Vector grammar, places, and the functional role of the spatial prepositions in English. In E. van der Zee & J. Slack (eds.), *Representing Direction in Language and Space.* Oxford: Oxford University Press, 69–85.

O'Regan, J. K. (1992) Solving the 'real' mysteries of visual perception: the world as an outside memory. *Canadian Journal of Psychology* 46: 461–88.

O'Regan, J. K. & Noë, A. (2001) A sensorimotor account of vision and visual consciousness. *Behavioral and Brain Sciences* 24: 939–1031.

O'Reilly, R. C., & Munakata, Y. (2000) *Computational Explorations in Cognitive Neuroscience: Understanding the Mind by Simulating the Brain.* Cambridge, MA: MIT Press.

Oakes, L. M. (1994) Development of infants' use of continuity cues in their perception of causality. *Developmental Psychology* 30: 869–79.

—— & Cohen, L. B. (1990) Infant perception of a causal event. *Cognitive Development* 5: 193–207.

Oates, T. (2001) Grounding knowledge in sensors. Ph.D. Dissertation, University of Massachusetts, Amherst.

Paivio, A., Yuille, J. C., & Smythe, P. C. (1966) Stimulus and response abstractness, imagery, and meaningfulness, and reported mediators in paired associate learning. *Canadian Journal of Psychology* 20: 362–77.

Palmer, S. E., & Hemenway, K. (1978) Orientation and symmetry: effects of multiple, rotational, and near symmetries. *Journal of Experimental Psychology: Human Perception and Performance* 4: 691–702.

Pasupathy A., & Connor, C. E. (2002) Population coding of shape in area V4. *Nature Neuroscience* 5: 1332–8.

Paterson, S. J., Brown, J. H., Gsödl, M. K., Johnson, M. H., & Karmiloff-Smith, A. (1999) Cognitive modularity and genetic disorders. *Science* 286: 2355–8.

Paul, B., Stiles, J., Passarotti, A., Bavar, N., & Bellugi, U. (2002) Face and place processing in Williams syndrome: evidence for a dorsal-ventral dissociation. *Neuroreport: For Rapid Communication of Neuroscience Research* 13: 1115–19.

Pecher, D., Zeelenberg, R., & Barsalou, L. W. (2003) Verifying properties from different modalities for concepts produces switching costs. *Psychological Science* 14: 119–24.

Pena, M., Bonatti, L. L., Nespor, M., & Mehler, J. (2002) Signal-driven computations in speech processing. *Science* 298(5593), 604–7.

Perky, C. W. (1910) An experimental study of imagination. *American Journal of Psychology* 21: 422–52.

Perrett, D. I., Oram, M. W., & Ashbridge, E. (1998) Evidence accumulation in cell populations responsive to faces: an account of generalisation of recognition without mental transformations. *Cognition* 67: 111–45.

Peterson, S. K., Mercer, C. D., & O'Shea, L. (1988) Teaching learning disabled students place value using the concrete to abstract sequence. *Learning Disabilities Research* 4: 52–6.

Piaget, J. (1951) *Play, Dreams, and Imitation in Childhood*. New York: Norton.

——(1954) *The Construction of Reality in the Child*. New York: Basic Books.

——(1963) *The Origins of Intelligence in Children*. New York: Norton.

——& Inhelder, B. (1956) *The Child's Conception of Space*. New York: Norton.

————(1967[1948]) *The Child's Conception of Space*, trans. F. J. Langdon & J. L. Lunzer. New York: Norton.

Pisoni, D. B., & Tash, J. (1974) Reaction times to comparisons within and across phonetic categories. *Perception and Psychophysics* 15: 285–90.

Plumert, J. M., & Hund, A. M. (2001) The development of location memory: what role do spatial prototypes play? *Child Development* 72: 370–84.

Plunkett, K., Sinha, C., Moller, M., & Strandsby, O. (1992) Symbol grounding or the emergence of symbols? Vocabulary growth in children and a connectionist net. *Connection Science* 4: 294–312.

Port, R. F., & van Gelder, T. (eds.) (1995) *Mind as Motion: Explorations in the Dynamics of Cognition*. Cambridge, MA: MIT Press.

Posner, M. I. (1980) Orienting of attention. *Quarterly Journal of Experimental Psychology* 32: 3–25.

Post, T. R. (1988) Some notes on the nature of mathematics learning. In T. R. Post (ed.), *Teaching Mathematics in Grades K–8: Research-Based Methods*. Boston, MA: Allyn & Bacon 1–19.

Pouget, A., Dayan, P., & Zemel, R. S. (2000) Inference and computation with population codes. *Annual Review of Neuroscience* 26: 381–410.

Premack, D. (1990) The infant's theory of self-propelled objects. *Cognition* 36: 1–16.

Pulverman, R., Hirsh-Pasek, K. & Golinkoff, R. M. (2006) Conceptual foundations for verb learning: celebrating the event. In K. Hirsh-Pasek & R. M. Golinkoff (eds.), *Action Meets Word: How Children Learn Verbs*. Oxford: Oxford University Press, 134–59.

Pulvermüller, F. (2002) *The Neuroscience of Language: On Brain Circuits of Words and Serial Order*. New York: Cambridge University Press.

Pylyshyn, Z. W. (1989) The role of location indexes in spatial perception: a sketch of the FINST spatial index model. *Cognition* 32: 65–97.

—— (2000) Situating the world in vision. *Trends in Cognitive Science* 4: 197–204.

—— (2001) Visual indexes, preconceptual objects, and situated vision. *Cognition* 80: 127–58.

—— & Storm, R. (1988) Tracking multiple independent objects: evidence for a parallel tracking mechanism. *Spatial Vision* 3: 179–97.

Quine, W. (1960) *Word and Object*. Cambridge, MA: MIT Press.

Rabiner, L. R., & Juang, B. (1989) A tutorial on hidden Markov models and selected applications in speech recognition. *Proceedings of the IEEE* 77: 257–86.

Radden, G. (1996) Motion metaphorized: the case of 'coming' and 'going'. In E. Casad (ed.), *Cognitive Linguistics in the Redwoods: The Expansion of a New Paradigm in Linguistics*. Berlin: Mouton de Gruyter, 423–58.

Rakison, D. H., & Poulin-Dubois, D. (2001) Developmental origin of the animate–inanimate distinction. *Psychological Bulletin* 127: 209–28.

Rao, R. P. N., Olshausen, B. A., & Lewicki, M. S. (eds.) (2002) *Probabilistic Models of the Brain: Perception and Neural Function*. Cambridge, MA: MIT Press.

Rattermann, M. J., & Gentner, D. (1998) The effect of language on similarity: the use of relational labels improves young children's performance in a mapping task. In K. Holyoak, D. Gentner, & B. Kokinov (eds.), *Advances in Analogy Research: Integration of Theory and Data from Cognitive, Computational, and Neural Sciences*. Sofia, Bulgaria: New Bulgarian University, 274–82.

Regier, T. (1996) *The Human Semantic Potential: Spatial Language and Constrained Connectionism*. Cambridge, MA: MIT Press.

—— & Carlson, L. A. (2001) Grounding spatial language in perception: an empirical and computational investigation. *Journal of Experimental Psychology: General* 130: 273–98.

Reimer, K., & Moyer, P. S. (2005) Third-graders learn about fractions using virtual manipulatives: a classroom study. *Journal of Computers in Mathematics and Science Teaching* 24: 5–21.

Reiss, J. E., Hoffman, J. E., & Landau, B. (2005) Motion processing specialization in Williams syndrome. *Vision Research* 45: 3379–90.

Remington, R. W., & Folk, C. (2001) A dissociation between attention and selection. *Psychological Science* 12: 511–15.

Resnick, L. B., & Omanson, S. F. (1987) Learning to understand arithmetic. In R. Glaser (ed.), *Advances in Instructional Psychology*, vol. 3. Hillsdale, NJ: Erlbaum, 41–95.

Richards, D., & Goldfarb, J. (1986) The episodic memory model of conceptual development: an integrative viewpoint. *Cognitive Development* 1: 183–219.

Richardson, D. C., & Kirkham, N. Z. (2004) Multimodal events and moving locations: eye movements of adults and 6-month-olds reveal dynamic spatial indexing. *Journal of Experimental Psychology: General* 133: 46–62.

—— —— (in press) Multimodal events and moving locations: evidence for dynamic spatial indexing in adults and six month olds. *Journal of Experimental Psychology: General*.

——— & Spivey, M. (2000) Representation, space, and Hollywood squares: looking at things that aren't there anymore. *Cognition* 76: 269–95.

——— ——— Barsalou, L. W., & McRae, K. (2003) Spatial representations activated during real-comprehension of verbs. *Cognitive Science* 27: 767–80.

——— ——— Edelman, S., & Naples, A. D. (2001) 'Language is spatial': experimental evidence for image schemas of concrete and abstract verbs. *Proceedings of the 23rd Annual Conference of the Cognitive Science Society*. Mahwah, NJ: Erlbaum, 845–50.

Rittle-Johnson, B., & Alibali, M. W. (1999) Conceptual and procedural knowledge of mathematics: does one lead to the other? *Journal of Educational Psychology* 91: 175–89.

Robinson, C. (1995) *Dynamical Systems: Stability, Symbolic Dynamics, and Chaos*. Ann Arbor, MI: CRC Press.

Robinson, D. N. (1986) Cognitive psychology and philosophy of mind. In T. Knapp and L. Robertson (eds.), *Approaches to Cognition: Contrasts and Controversies*. Hillsdale, NJ: Erlbaum, 1–11.

Roennberg, J. (1990) On the distinction between perception and cognition. *Scandinavian Journal of Psychology* 31: 154–6.

Rogoff, B., & Chavajay, P. (1995) What's become of research on the cultural basis of cognitive development? *American Psychologist* 50: 859–77.

Rolls, E. T., & Tovee, M. J. (1995) Sparseness of the neuronal representation of stimuli in the primate temporal visual cortex. *Journal of Neurophysiology* 73: 713–26.

Rose, D. (1996) Some reflections on (or by?) grandmother cells. *Perception* 25: 881–6.

Rosen, R. (2000) *Essays on Life Itself*. New York: Columbia University Press.

Roy, D., & Pentland, A. (2002) Learning words from sights and sounds: a computational model. *Cognitive Science* 26: 113–46.

Saffran, J. R., Johnson, E., Aslin, R., & Newport, E. (1999) Statistical learning of tone sequences by human infants and adults. *Cognition* 70: 27–52.

——— Newport, E. L., & Aslin, R. N. (1996) Word segmentation: the role of distributional cues. *Journal of Memory and Language* 35: 606–21.

Salvucci, D. D., & Goldberg, J. H. (2000) Identifying fixations and saccades in eye-tracking protocols. In *Proceedings of Eye Tracking Research and Applications Symposium*. New York: ACM Press, 71–8.

Samuelson, L. K., & Smith, L. B. (1998) Memory and attention make smart word learning: an alternative account of Akhtar, Carpenter and Tomasello. *Cognitive Development* 1: 94–104.

Samuelson, S., & Smith, L. (2000) Grounding development in cognitive processes. *Child Development* 71: 98–106.

Sandhofer, CJU. & Smith, L.B. (1999). Learning color words involves learning a system of mappings. *Developmental Psychology*, 35, 668–679.

Santa, J. L. (1977) Spatial transformations of words and pictures. *Journal of Experimental Psychology: Human Learning and Memory* 3: 418–27.

Schaeffer, B., Eggleston, V. H., & Scott, J. L. (1974) Number development in young children. *Cognitive Psychology* 6: 357–79.

Schirra, J. (1993) A contribution to reference semantics of spatial prepositions: the visualization problem and its solution in VITRA. In C. Zelinsky-Wibbelt (ed.), *The Semantics of Prepositions: From Mental Processing to Natural Language Processing.* Berlin: Mouton de Gruyter, 471–515.

Schober, M. F. (1995) Speakers, addressees, and frames of reference: whose effort is minimized in conversations about locations? *Discourse Processes* 20: 219–47.

Schoenfeld, A. H. (1987) When good teaching leads to bad results: the disasters of 'well taught' mathematics courses. In P. L. Peterson & T. L. Carpenter (eds.), *Educational Psychologist* 23: 145–66.

Scholl, B. J., & Pylyshyn, Z. W. (1999) Tracking multiple items through occlusion: clues to visual objecthood. *Cognitive Psychology* 38: 259–90.

Schutte, A. R., & Spencer, J. P. (2002) Generalizing the dynamic field theory of the A-not-B error beyond infancy: three-year-olds' delay- and experience-dependent location memory biases. *Child Development* 73: 377–404.

——— (2007) Planning 'discrete' movements using a continuous system: insights from a dynamic field theory of movement preparation. *Motor Control* 11: 166–208.

Schutte, A. R., & Spencer, J. P. (in press) Tests of the dynamic field theory and the spatial precision hypothesis: capturing a qualitative developmental transition in spatial working memory. *Journal of Experimental Psychology: Human Perception and Performance.*

——— & Schöner, G. (2003) Testing the dynamic field theory: working memory for locations becomes more spatially precise over development. *Child Development* 74: 1393–1417.

Scripture, E. W. (1896) Measuring hallucinations. *Science* 3: 762–3.

Searle, J. R. (1980) Minds, brains, and programs. *Brain and Behavioral Sciences* 3: 417–57.

Segal, S., & Gordon, P. E. (1969) The Perky Effect revisited: blocking of visual signals by imagery. *Perceptual and Motor Skills* 28: 791–7.

Shallice, T. (1996) The language-to-object perception interface: evidence from neuropsychology. In P. Bloom, M. A. Peterson, L. Nadel, & M. F. Garrett (eds.), *Language and Space.* Cambridge, MA: MIT Press, 531–52.

Shelton, A. L., & McNamara, T. P. (2001) Systems of spatial reference in human memory. *Cognitive Psychology* 43: 274–310.

Shepard, R. N. (1975) Form, formation, and transformation of internal representations. In R. L. Solso (ed.), *Information Processing and Cognition: The Loyola Symposium.* Hillsdale, NJ: Erlbaum, 87–122.

——— & Chipman, S. (1970) Second-order isomorphism of internal representations: shapes of states. *Cognitive Psychology* 1: 1–17.

——— & Hurwitz, S. (1984) Upward direction, mental rotation, and discrimination of left and right turns in maps. Special issue on visual cognition, *Cognition* 18(1–3): 161–93.

——— & Metzler, J. (1971) Mental rotation of three-dimensional objects. *Science* 171: 701–3.

Siegler, R. S., & Robinson, M. (1982) The development of numerical understandings. In H. W. Reese & L. P. Lipsitt (eds.), *Advances in Children Development and Behavior.* New York: Academic Press. 243–312.

Simmering, V. R., Schutte, A. R., & Spencer, J. P. (2008) Generalizing the dynamic field theory of spatial working memory across real and developmental time scales. In S. Becker (ed.), *Computational Cognitive Neuroscience*, special issue of *Brain Research* 1202: 68–86.

Simon, M. A. (1995). Reconstructing mathematics pedagogy from a constructivist perspective. *Journal for Research in Mathematics Education*, 26: 114–145.

Simmons, W. K., and Barsalou, L. W. (2003) The similarity-in-topography principle: reconciling theories of conceptual deficits. *Cognitive Neuropsychology* 20: 451–86.

Simons, D. J., & Wang, R. F. (1998) Perceiving real-world viewpoint changes. *Psychological Science* 9: 315–20.

Simos, P. G., Diehl, R. L., Breier, J. I., Molis, M. R., Zouridakis, G., & Papanicolaou, A. C. (1998) MEG correlates of categorical perception of a voice onset time continuum in humans. *Cognitive Brain Research* 7: 215–19.

Siskind, J. M. (1996) A computational study of cross-situational techniques for learning word-to-meaning mappings. *Cognition* 61: 39–61.

Slobin, D. I. (1973) Cognitive prerequisites for the development of grammar. In C. A. Ferguson & D. I. Slobin (eds.), *Studies in Child Language Development*. New York: Holt, Rinehart & Winston, 175–208.

Sloman, S. A. (1996) The empirical case for two systems of reasoning. *Psychological Bulletin* 119: 3–22.

Smith, L. B. (1993) The concept of same. In H. W. Reese (ed.), *Advances in Child Development and Behavior*, vol. 24. New York: Academic Press 215–52.

—— (2000) How to learn words: an associative crane. In R. Golinkoff & K. Hirsh-Pasek (eds.), *Breaking the Word Learning Barrier*. Oxford: Oxford University Press, 51–80.

—— (2005) Action alters shape categories. *Cognitive Science* 29: 665–79.

—— (2009) Dynamic systems, sensori-motor processes, and the origins of stability and flexibility. In J. P. Spencer, M. Thomas, & J. McClelland (eds.), *Toward a Unified Theory of Development: Connectionism and Dynamic Systems Theory Reconsidered*. New York: Oxford University Press, 67–85.

—— Clearfield, M., Diedrich, F., & Thelen, E. (in preparation) Evidence for embodiment: memory tied to the body's current position.

—— Jones, S. S., & Landau, B. (1996) Naming in young children: a dumb attentional mechanism? *Cognition* 60: 143–71.

—— Thelen, E., Titzer, R., & McLin, D. (1999) Knowing in the context of acting: the task dynamics of the A-not-B error. *Psychological Review* 106: 235–60.

Solomon, K. O., & Barsalou, L. W. (2001) Representing properties locally. *Cognitive Psychology* 43: 129–69.

Sowell, E. J. (1989) Effects of manipulative materials in mathematics instruction. *Journal for Research in Mathematics Education* 20: 498–505.

Sparks, D. L., Holland, R., & Guthrie, B. L. (1976) Size and distribution of movement fields in the monkey superior colliculus. *Brain Research* 113: 21–34.

Spelke, E. (2003) What makes us smart? Core knowledge and natural language. In D. Gentner & S. Goldin-Meadow (eds.), *Language in Mind*. Cambridge, MA: MIT Press, 277–311.

Spelke, E., & Hespos, S. (2001) Continuity, competence, and the object concept. In E. Dupoux (ed.), *Language, Brain, and Cognitive Development: Essays in honor of Jacques Mehler.* Cambridge, MA: MIT Press, 325–40.

——Phillips, A. T., & Woodward, A. L. (1995) Infants' knowledge of object motion and human action. In D. Sperber, D. Premack, & A. Premack (eds.), *Causal Cognition: A Multidisciplinary Debate.* New York: Oxford University Press, 44–78.

——& Tsivkin, S. (2001) Innate knowledge and conceptual change: space and number. In M. Bowerman & S. C. Levinson (eds.), *Language Acquisition and Conceptual Development.* New York: Cambridge University Press, 70–97.

Spencer, J. P., & Hund, A. M. (2002) Prototypes and particulars: geometric and experience-dependent spatial categories. *Journal of Experimental Psychology: General* 131: 16–37.

Spencer, J. P., & Hund, A. M. (2003) Developmental continuity in the processes that underlie spatial recall. *Cognitive Psychology* 47: 432–80.

——& Schöner, G. (2003) Bridging the representational gap in the dynamic systems approach to development. *Developmental Science* 6: 392–412.

——Simmering, V. R., Schutte, A. R., & Schöner, G. (2007) What does theoretical neuroscience have to offer the study of behavioral development? Insights from a dynamic field theory of spatial cognition. In J. Plumert & J. P. Spencer (eds.), *The Emerging Spatial Mind.* Oxford: Oxford University Press, 320–61.

——Smith, L. B., & Thelen, E. (2001) Tests of a dynamic systems account of the A-not-B error: the influence of prior experience on the spatial memory abilities of two-year-olds. *Child Development* 72: 1327–46.

Spivey, M. J. (2007) *The Continuity of Mind.* New York: Oxford University Press.

——& Geng, J. J. (2001) Oculomotor mechanisms activated by imagery and memory: eye movements to absent objects. *Psychological Research* 65: 235–41.

——Richardson, D. C., & Fitneva, S. A. (2004) Thinking outside the brain: spatial indices to linguistic and visual information. In J. Henderson and F. Ferreira (eds.), *The Interaction of Vision, Language, and Action.* New York: Psychology Press, 161–89.

————& Gonzalez-Marquez, M. (2005) On the spatial and image schematic underpinnings of real-time language processing. In D. Pecher & R. Zwaan (eds.), *The Grounding of Cognition: The Role of Perception and Action in Memory, Language, and Thinking.* Cambridge: Cambridge University Press, 246–81.

——Tanenhaus, M., Eberhard, K., & Sedivy, J. (2002) Eye movements and spoken language comprehension: effects of visual context on syntactic ambiguity resolution. *Cognitive Psychology* 45: 447–81.

————Eberhard, K., & Tanenhaus, M. (2001) Linguistically mediated visual search. *Psychological Science* 12: 282–6.

——Tyler, M., Richardson, D. C., & Young, E. E. (2000) Eye movements during comprehension of spoken scene descriptions. *Proceedings of the 22nd Annual Conference of the Cognitive Science Society.* Mahwah, NJ: Erlbaum, 487–92.

Spivey-Knowlton, M. J., Tanenhaus, M. K., Eberhard, K. M., & Sedivy, J. C. (1998) Integration of visuospatial and linguistic information: language comprehension in real time and real space. In P. Oliver & K. P. Gapp (eds.), *Representation and Processing of Spatial Expressions*. Mahwah, NJ: Erlbaum, 201–14.

Stanfield, R. A., & Zwaan, R. A. (2001) The effect of implied orientation derived from verbal context on picture recognition. *Psychological Science* 12: 153–6.

Stein, J. F. (1992) The representation of egocentric space in the posterior parietal cortex. *Behavioral and Brain Sciences* 15: 691–700.

Steinschneider, M., Schroeder, C. E., Arezzo, J. C., & Vaughan, H. G. (1995) Physiologic correlates of the voice onset time boundary in primary auditory cortex (A1) of the awake monkey: temporal response patterns. *Brain and Language* 48: 326–40.

Stevenson, H. W., & Stigler, J. W. (1992) *The Learning Gap: Why Our Schools Are Failing and What We Can Learn from Japanese and Chinese Education*. New York: Simon & Schuster.

Suydam, M. N. (1986) Manipulative materials and achievement. *Arithmetic Teacher* 33: 10–32.

—— & Higgins, J. L. (1977) *Activity-Based Learning in Elementary School Mathematics: Recommendations from Research*. Columbus, OH: ERIC Center for Science, Mathematics, and Environmental Education, College of Education, Ohio State University.

Sweetser, E. E. (1990) *From Etymology to Pragmatics: The Mind-as-Body Metaphor in Semantic Structure and Semantic Change*. Cambridge: Cambridge University Press.

Swindale, N. (2001) Cortical cartography: what's in a map? *Current Biology* 11: R764–7.

Tager-Flusberg, H., Plesa-Skwerer, D., & Faja, S. (2003) People with Williams syndrome process faces holistically. *Cognition* 89: 11–24.

Tallal, P., Galaburda, A. M., Llinás, R. R., & von Euler, C. (eds.) (1993) *Temporal Information Processing in the Nervous System: Special Reference to Dyslexia and Dysphasia*. New York: New York Academy of Sciences.

Talmy, L. (1975) Semantics and syntax of motion. In J. Kimball (ed.), *Syntax and Semantics 4*. New York: Academic Press, 181–238.

—— (1983) How language structures space. In H. L. Pick & L. P. Acredolo (eds.), *Spatial Orientation: Theory, Research and Application*. New York: Plenum Press, 225–82.

—— (1988) Force dynamics in language and cognition. *Cognitive Science* 12: 49–100.

—— (1996) Fictive motion in language and 'ception'. In P. Bloom, M. A. Peterson, L. Nadel, & M. F. Garrett (eds.), *Language and Space*. Cambridge, MA: MIT Press, 211–76.

—— (2000) *Towards a Cognitive Semantics*, vol. 1: *Conceptual Structuring Systems*. Cambridge: MIT Press.

Tanaka, K. (1997) Mechanisms of visual object recognition: monkey and human studies. *Current Opinion in Neurobiology* 7: 523–9.

—— Spivey-Knowlton, M. J., Eberhard, K. M., & Sedivy, J. C. (1995) Integration of visual and linguistic information in spoken language comprehension. *Science* 268: 1632–4.

Tanenhaus, M., & Trueswell, J. (1995) Sentence comprehension. In J. Miller & P. Eimas (eds.), *Handbook of Cognition and Perception*. New York: Academic Press, 217–62.

Tardif, T. (1996) Nouns are not always learned before verbs: evidence from Mandarin speakers' early vocabularies. *Developmental Psychology* 32: 492–504.

Tardif, T., Shatz, M., & Naigles, L. (1997) Caregiver speech and children's use of nouns versus verbs: a comparison of English, Italian, and Mandarin. *Journal of Child Language* 24: 535–65.

Tenenbaum, J., & Xu, F. (2000) Word learning as Bayesian inference. In L. Gleitman & A. Joshi (eds.), *Proceedings of the 22nd Annual Conference of the Cognitive Science Society.* Mahway, NJ: Erlbaum, 517–22.

Thelen, E. (1995) Time-scale dynamics and the development of an embodied cognition. In R. F. Port & T. van Gelder (eds.), *Mind as Motion: Explorations in the Dynamics of Cognition.* Cambridge, MA: MIT Press, 69–100.

——Schöner, G., Scheier, C., & Smith, L. B. (2001) The dynamics of embodiment: a field theory of infant perseverative reaching. *Behavioral and Brain Sciences* 24: 1–86.

——& Smith, L. B. (1994) *A Dynamic Systems Approach to the Development of Cognition and Action.* Cambridge: MIT Press.

Thelen, E. & Smith, L. B. (1998) Dynamic systems theories. In R. M. Lerner (ed.), *Theoretical Models of Human Development,* 5th edn., vol. 1. New York: Wiley, 563–634.

Thompson, P. W. (1992) Notions, conventions, and constraints: contributions to effective uses of concrete materials in elementary mathematics. *Journal for Research in Mathematics Education* 23: 123–47.

——(1994) Concrete materials and teaching for mathematical understanding. *Arithmetic Teacher* 41: 556–8.

Thompson, R. K. R., Oden, D. L., & Boysen, S. T. (1997) Language-naive chimpanzees (*Pan troglodytes*) judge relations between relations in a conceptual matching-to-sample task. *Journal of Experimental Psychology: Animal Behavior Processes* 23: 31–43.

Tipper, S. P., & Behrmann, M. (1996) Object centered not scene based visual neglect. *Journal of Experimental Psychology: Human Perception and Performance* 22: 1261–78.

Tolman, E. C. (1948) Cognitive maps in rats and men. *Psychological Review* 55: 189–208.

Tomasello, M. (1992) *First Verbs: A Case Study of Early Grammatical Development.* Cambridge: Cambridge University Press.

——(1995) Pragmatic contexts for early verb learning, In M. Tomasello & W. E. Merriman (eds.), *Beyond Names for Things: Young Children's Acquisition of Verbs.* Hillsdale, NJ: Erlbaum, 115–46.

——(2000) Perceiving intentions and learning words in the second year of life. In M. Bowerman & S. Levinson (eds.), *Language Acquisition and Conceptual Development.* Cambridge: Cambridge University Press, 111–28.

——& Farrar, M. (1986) Joint attention and early language. *Child Development* 57: 1454–63.

Toskos, A., Hanania, R., & Hockema, S. (2004) Eye scanpaths influence memory for spoken verbs. In *Proceedings of the 26th Annual Conference of the Cognitive Science Society,* 1643.

Tovee, M. J., & Rolls, E. T. (1992) The functional nature of neuronal oscillations. *Trends in Neurosciences* 15: 387.

Towell, G. G., & Shavlik, J. W. (1993) Extracting refined rules from knowledge based neural networks. *Machine Learning* 13: 71–101.

Trappenberg, T. P., Dorris, M. C., Munoz, D. P., & Klein, R. M. (2001) A model of saccade initiation based on the competitive integration of exogenous and endogenous signals in the superior colliculus. *Journal of Cognitive Neuroscience* 13: 256–71.

Tversky, B. (1991) Spatial mental models. In G. Bower (ed.), *Psychology of Learning and Motivation*. San Diego, CA: Academic Press, 109–45.

Udwin, O., Davies, M., & Howlin, P. (1996) A longitudinal study of cognitive abilities and educational attainment in Williams syndrome. *Developmental Medicine and Child Neurology* 38: 1020–29.

Ullman, S. (1984) Visual routines. Special issue on visual cognition, *Cognition* 18(1–3): 97–159.

Uttal, D. H., Liu, L. L., & Deloache, J. S. (2006) Concreteness and symbolic development. In L. Balter & C. S. Tamis-Lemonda (eds.), *Child Psychology: Handbook of Contemporary Issues*, 2nd edn. New York: Psychology Press, 167–84.

——Scudder, K. V., & Deloache, J. S. (1997) Manipulatives as symbols: a new perspective on the use of concrete objects to teach mathematics. *Journal of Applied Developmental Psychology* 18: 37–54.

Vallortigara, G., Zanforlin, M., & Pasti, G. (1990) Geometric modules in animals' spatial representations: a test with chicks (*Gallus gallus*). *Journal of Comparative Psychology* 104: 248–54.

van der Zee, Emile (1996) Spatial knowledge and spatial language. Dissertation, ISOR/Utrecht University.

Van Orden, G. C., Holden, J. G., & Turvey, M. T. (2003) Self organization of cognitive performance. *Journal of Experimental Psychology: General* 132: 331–50.

Vandeloise, C. (1991) *Spatial Prepositions: A Case Study from French*. Chicago, IL: University of Chicago Press.

vanMarle, K., & Scholl, B. (2003) Attentive tracking of objects versus substances. *Psychological Science* 14: 498–504.

von der Malsburg, C. (1973) Self-organization of orientation sensitive cells in the striate cortex. *Kybernetik* 14: 85–100.

Von Neumann, J. (1958) *The Computer and the Brain*. New Haven, CT: Yale University Press.

Vygotsky, L. (1978) *Mind in Society: The Development of Higher Psychological Processes*. Cambridge, MA: Harvard University Press.

——(1986) *Thought and Language*. Cambridge, MA: MIT Press.

Wagner, S., & Walters, J. A. (1982) A longitudinal analysis of early number concepts: from numbers to number. In G. Forman (ed.), *Action and Thought*. New York: Academic Press 137–61.

Wang, R. F., & Spelke, E. (2002) Human spatial representation: insights from animals. *Trends in Cognitive Sciences* 6: 376–82.

Ward, L. (2002) *Dynamical Cognitive Science*. Cambridge, MA: MIT Press.

Waxman, S. R. & Hall, D. G. (1993). The development of a linkage between count nouns and object categories: Evidence from fifteen- to twenty-one-monthold infants. *Child Development* 64: 1224–1241.

Waxman, S. R., & Markow, D. B. (1995) Words as invitations to form categories: Evidence from 12- to 13-month-old infants. *Cognitive Psychology* 29: 257–302.

Wearne, D., & Hiebert, J. (1988) A cognitive approach to meaningful mathematics instruction: testing a local theory using decimal numbers. *Journal for Research in Mathematics Education* 19: 371–84.

Wellman, H. M. (1986) *Infant Search and Object Permanence: A Meta-Analysis of the A-Not-B Error.* Ann Arbor, MI: Society for Research in Child Development.

Wenderoth, P., & van der Zwan, R. (1991) Local and gloabal mechanisms of one- and two-dimensional orientation illusions. *Perception and Psychophysics* 50: 321–32.

Wickelgren, W. A. (1977) Concept neurons: a proposed developmental study. *Bulletin of the Psychonomic Society* 10: 232–4.

Woodward, A. L. (1998) Infants selectively encode the goal object of an actor's reach. *Cognition* 69: 1–34.

——& Guajardo, J. (2002) Infants' understanding of the point gesture as an object-directed action. *Cognitive Development* 17: 1061–84.

Wraga, M., Creem, S. H., & Proffitt, D. R. (2000) Updating displays after imagined object and viewer rotations. *Journal of Experimental Psychology: Learning, Memory, and Cognition* 26: 151–68.

Wynn, K. (1990) Children's understanding of counting. *Cognition* 36: 155–93.

——(1992) Children's acquisition of the number words and the counting system. *Cognitive Psychology* 24: 220–51.

——(1998) Psychological foundations of number: numerical competence in human infants. *Trends in Cognitive Science* 2: 296–303.

Young, M. P., & Yamane, S. (1992) Sparse population coding of faces in the inferotemporal cortex. *Science* 256: 1327–31.

Yu, C., & Ballard, D. H. (2004) A multimodal learning interface for grounding spoken language in sensory perceptions. *ACM Transactions on Applied Perception* 1: 57–80.

——————& Aslin, R. N. (2005) The role of embodied intention in early lexical acquisition. *Cognitive Science* 29: 961–1005.

Zacks, J. M. (2004) Using movement and intention to explain simple events. *Cognitive Science* 28: 979–1008.

Zheng, M., & Goldin-Meadow, S. (2002) Thought before language: how deaf and hearing children express motion events across cultures. *Cognition* 85: 145–75.

Zwaan, R. A. (1999) Situation models: the mental leap into imagined worlds. *Current Directions in Psychological Science* 8: 15–18.

——(2004) The immersed experiencer: toward an embodied theory of language comprehension. In B. H. Ross (ed.), *The Psychology of Learning and Motivation*, vol. 44. New York: Academic Press, 35–62.

——Stanfield, R. A., & Yaxley, R. H. (2002) Do language comprehenders routinely represent the shapes of objects? *Psychological Science* 13: 168–71.

——Madden, C. J., Yaxley, R. H., & Aveyard, M. E. (2004) Moving words: dynamic representations in language comprehension. *Cognitive Science* 28: 611–19.

# Author Index

# Subject Index